KEENE VALLEY REGION WATERFALL GUIDE

The Search for Cool Cascades
in the Heart of the Adirondacks

North Hudson to Keeseville, St. Huberts to Lake Placid

Russell Dunn

Published by Black Dome Press Corp.
649 Delaware Ave., Delmar, N.Y. 12054
blackdomepress.com
(518) 439-6512

First Edition Paperback 2017
Copyright © 2017 by C. Russell Dunn

Without limiting the rights under copyright above, no part of this publication may be reproduced, stored in or introduced into a retrieval system, or transmitted, in any form, by any means (electronic, mechanical, photocopying, recording, or otherwise), without the prior written permission of the publisher of this book.

ISBN: 978-1-883789-87-9

Library of Congress Control Number: 2017956757

CAUTION: Outdoor recreational activities are by their very nature potentially hazardous and contain risk. See Caution: Safety Tips, p xvii.

Front cover: *Roaring Brook Falls* by John Haywood
Back cover, left to right: *Split Rock Falls* by John Haywood, *Beer Walls Falls* by John Haywood, *Fairy Ladder Falls* by John Holmes

Maps created by John Haywood, using USGS map software.
Photographs by author unless otherwise noted.

Printed in the USA

10 9 8 7 6 5 4 3 2 1

To Christy Butler, fellow author and photographer, who has done more than his share to popularize waterfalls in the Northeast United States.

"Waterfalls are unique landforms, far rarer than mountains or valleys, glaciers or lakes. Even the smallest waterfall surprises and delights visitors, but viewing a waterfall is also an opportunity to see geology interact with hydrology as a soft material (in this case, water) sculpts a hard material such as rock."
Geoffrey H. Nash, *The Extreme Earth: Waterfalls*

"It's a great thing these days to leave civilization for a while."
Robert Marshall, environmental activist, founder of the *Wilderness Society*, and one of the three original Adirondack Forty-Sixers.

Contents

Regional Maps	x
Foreword by Neal Burdick	xv
Caution: Safety Tips	xvii
Mileage	xxiii
Introduction	xxiv
Why write about waterfalls?	xxv
Why do waterfalls exist?	xxix
The Adirondacks	xxxi
Mister Indestructible?	xxxiii
John Haywood's Tips on Photographing Waterfalls	xxxvi

Section I: The Boquet   1

1. Split Rock Falls	3
2. Underwood Falls	7
Falls North of Split Rock Falls	8
3. North Fork Gorge Cascades	8
4. Cascade on Twin Pond Outlet Stream	10
5. Shoebox Falls	11
6. Upper Boquet Falls	13
7. Boquet Canyon & Falls	14
8. Southside Approach to River & Rapids	16
9. Chasm Cascade	17
10. Interim Stop	19
11. Falls on South Fork of the Boquet	20
Falls South of Split Rock Falls.	21
12. Cascades along East Trail to Giant Mountain	22
13. Fall in New Russia	22
14. Rice's Falls and US Gorge	23
15. Silver Cascade	26
16. Falls near Knob Lock Mountain	28
17. Wadhams Falls	28
18. Merriam Forge Falls	32
19. Willsboro Falls	34

Section II: North Hudson   37

20. Blue Ridge Falls	38

21. The Falls	39
22. West Mill Brook Falls	41
23. Lindsay Brook Falls	44
24. Crowfoot Brook Cascades	46
25. Ash Craft Brook Falls	48

Section III: Chapel Pond Pass 50

26. Hidden Cascade	51
27. Secluded Cascade	52
28. Twin Pond Cascade	54
29. Twin Falls	56
30. Rock Garden Falls	58
31. Chapel Pond Slab Cascade & Mini Trap Dike	59
32. Beede Brook Falls	60
33. Chapel Pond Falls	65
34. Beer Walls Falls	67
35. Roaring Brook Falls	70

Section IV: Adirondack Mountain Reserve 74

Falls along the West River Trail	78
36. Pyramid Falls	79
37. First Falls on East Branch of Ausable River	80
38. Wedge Brook Falls	80
39. Second Falls on East Branch of Ausable River	82
40. Beaver Meadow Falls	82
41. Rainbow Falls	84
42. Falls along the East River Trail	89
43. Russell Falls	90
Falls on Gill Brook	91
44. The Flume	93
45. Gill Brook Steps	94
46. Artist Falls	94
47. More Cascades	96
48. Split Falls	96
49. Unnamed Scenic Cascades	98
50. Upper Gill Brook Falls	99
51. Fairy Ladder Falls	101

Section V: Keene Valley Area — 104

- 52. Mossy Cascade — 105
- 53. Deer Brook Falls and Deer Brook Gorge Cascades — 109
- 54. Roadside Falls — 112
- 55. Flume Falls — 114
- 56. Hopkins Brook Falls — 115
- 57. Phelps Falls — 117
- 58. Town Ridge Loop Trail Falls — 119
- 59. Blueberry Falls — 121
- 60. Hulls Falls — 122
- 61. Hull Basin Brook Falls — 124
- 62. Champagne Falls and Gristmill Road Gorge — 125

Section VI: Johns Brook — 127

Phelps Trail — 129
- 63. Deer Brook Cascades — 129
- 64. Upper Johns Brook Flume & Falls — 131
- 65. Slide Mountain Brook Falls — 132
  - *Johns Brook Lodge* — 133
- 66. Minor Cascades — 135
- 67. Bushnell Falls — 135

Southside Trail — 137
- 68. Rock Cut Brook Falls — 141
- 69. Funnel Falls — 141
- 70. Tenderfoot Pool Falls — 142
- 71. Johns Brook Lower Flume — 143
- 72. Wolf Jaw Brook Falls — 144

Section VII: From Keene to Keeseville — 147

- 73. Styles Brook Falls — 148
- 74. Jay Falls — 149
  - *Ausable Forks* — 151
- 75. Anderson Falls — 151
- 76. Indian Falls — 153
- 77. Alice Falls — 154
- 78. Dammed Waterfall — 156
  - *Ausable Chasm* — 157
- 79. Rainbow Falls — 158
- 80. Horseshoe Falls — 161

81. Tiny Cascades and Rapids	162
82. Buttermilk Falls	163

Section VIII: From Keene to Heart Lake — 164

83. Nichols Brook Falls	165
84. Clifford Falls	167
85. Cascade Lake Falls	168
Adirondack Loj / Heart Lake Area	171
86. Klondike Brook Falls	173
87. Rocky Falls	176
88. MacIntyre Falls	178
89. Indian Falls	179
90. Avalanche Pass Falls	182
91. Trap Dike Falls	185
92. Falls on Tributary to Lake Colden	187

Section IX: Northville-Placid Trail — 190

93. Chubb River Chasm	192
94. Minor Cascades on the Chubb River	193
95. Chubb River Falls	193
96. Wanika Falls	195

Section X: Wilmington Notch — 199

97. Whiteface Brook Falls	200
98. Holcomb Pond Cascade	204
99. Quarry Pool Falls	204
100. Monument Falls	206
101. Owen Pond's Outlet Cascade	208
102. Interim Point of Interest	209
103. High Falls Gorge	210
104. Wilmington Notch Falls from Wilmington Notch Campground	212
Falls at Whiteface Mountain Ski Center	213
105. Wilmington Notch Falls	215
106. Stag Brook Falls & Upper Cascades	216
107. Cascade on Stream Paralleling Stag Brook	220
Falls near Wilmington	221
108. Flume Falls	221
109. Whiteface Veteran's Memorial Highway Cascade	225
110. French's Brook Falls	226

Postscript 230
Acknowledgments 231
About the Author 232
Index 234

Upper Johns Brook Flume.

Regional Maps

1. Split Rock Falls (Map 1)
2. Underwood Falls (Map 1)
3. North Fork Gorge Cascades (Map 1)
4. Cascade on Twin Pond Outlet Stream (Map 1)
5. Shoebox Falls (Map 1)
6. Upper Boquet Falls (Map 1)
7. Boquet Canyon & Falls (Map 1)
8. Southside Approach to River & Rapids (Map 1)
9. Chasm Cascade (Map 1)
10. Interim Stop (Map 1)
11. Falls on South Fork of the Boquet (Map 1)
12. Cascades along East Trail to Giant Mountain (Map 1)
13. Fall in New Russia (Map 1)
14. Rice's Falls & US Gorge (Map 2)
15. Silver Cascade (Map 2)
16. Falls near Knob Lock Mountain (Map 1)
17. Wadhams Falls (Map 2)
18. Merriam Forge Falls (Map 2)
19. Willsboro Falls (Map 2)
20. Blue Ridge Falls (Map 1)
21. The Falls (Maps 1, 3)
22. West Mill Brook Falls (Map 1)
23. Lindsay Brook Falls (Map 1)
24. Crowfoot Brook Cascades (Map 1)
25. Ash Craft Brook Falls (Map 1)
26. Hidden Cascade (Map 1)
27. Secluded Cascade (Map 1)
28. Twin Pond Cascade (Map 1)
29. Twin Falls (Map 1)
30. Rock Garden Falls (Map 1)
31. Chapel Pond Slab Cascade & Mini Trap Dike (Map 1)
32. Beede Brook Falls (Map 1)
33. Chapel Pond Falls (Map 1)
34. Beer Walls Falls (Map 1)
35. Roaring Brook Falls (Map 1)
36. Pyramid Falls (Maps 1, 3)
37. First Falls on East Branch of Ausable River (Maps 1, 3)
38. Wedge Brook Falls (Maps 1, 3)
39. Second Falls on East Branch of Ausable River (Maps 1, 3)
40. Beaver Meadow Falls (Maps 1, 3)
41. Rainbow Falls (Maps 1, 3)
42. Falls along the East River Trail (Maps 1, 3)
43. Russell Falls (Maps 1, 3)
44. The Flume (Maps 1, 3)
45. Gill Brook Steps (Maps 1, 3)
46. Artist Falls (Maps 1, 3)
47. More Cascades (Maps 1, 3)
48. Split Falls (Maps 1, 3)
49. Unnamed Scenic Cascades (Maps 1, 3)
50. Upper Gill Brook Falls (Maps 1, 3)
51. Fairy Ladder Falls (Maps 1, 3)
52. Mossy Cascade (Maps 1, 3)
53. Deer Brook Falls and Deer Brook Gorge Cascades (Maps 1, 3)
54. Roadside Falls (Maps 1, 3)
55. Flume Falls (Maps 1, 3)
56. Hopkins Brook Falls (Maps 1, 3)

57. Phelps Falls (Maps 1, 3)
58. Town Ridge Loop Trail Falls (Map 3)
59. Blueberry Falls (Map 3)
60. Hulls Falls (Map 3)
61. Hull Basin Brook Falls (Map 3)
62. Champagne Falls and Gristmill Road Gorge (Map 3)
63. Deer Brook Cascade (Maps 1, 3)
64. Upper Johns Brook Flume & Falls (Map 3)
65. Slide Mountain Brook Falls (Map 3)
66. Minor Cascades (Map 3)
67. Bushnell Falls (Map 3)
68. Rock Cut Brook Falls (Maps 1, 3)
69. Funnel Falls (Maps 1, 3)
70. Tenderfoot Pool Falls (Maps 1, 3)
71. Johns Brook Lower Flume (Map 3)
72. Wolf Jaw Brook Falls (Map 3)
73. Styles Brook Falls (Map 3)
74. Jay Falls (Map 3)
75. Anderson Falls (Map 2)
76. Indian Falls (Map 2)
77. Alice Falls (Map 2)
78. Dammed Waterfall (Map 2)
79. Rainbow Falls (Map 2) (Map 2)
80. Horseshoe Falls (Map 2)
81. Tiny Cascades and Rapids (Map 2)
82. Buttermilk Falls (Map 2)
83. Nichols Brook Falls (Map 3)
84. Clifford Falls (Map 3)
85. Cascade Lake Falls (Map 3)
86. Klondike Brook Falls (Map 3)
87. Rocky Falls (Map 3)
88. MacIntyre Falls (Map 3)
89. Indian Falls (Map 3)
90. Avalanche Pass Falls (Map 3)
91. Trap Dike Falls (Map 3)
92. Falls on Tributary to Lake Colden (Map 3)
93. Chubb River Chasm (Map 3)
94. Minor Cascades on the Chubb River (Map 3)
95. Chubb River Falls (Map 3)
96. Wanika Falls (Map 3)
97. Whiteface Brook Falls (Map 3)
98. Holcomb Pond Cascade (Map 3)
99. Quarry Pool Falls (Map 3)
100. Monument Falls (Map 3)
101. Owen Pond's Outlet Cascades (Map 3)
102. Interim Point of Interest (Map 3)
103. High Falls Gorge (Map 3)
104. Wilmington Notch Falls from Wilmington Notch Campground (Map 3)
105. Wilmington Notch Falls (Map 3)
106. Stag Brook Falls & Upper Cascades (Map 3)
107. Cascade on Stream Paralleling Stag Brook (Map 3)
108. Flume Falls (Map 3)
109. Whiteface Veteran's Memorial Highway Cascade (Map 3)
110. French's Brook Falls (Map 3)

MAP 1

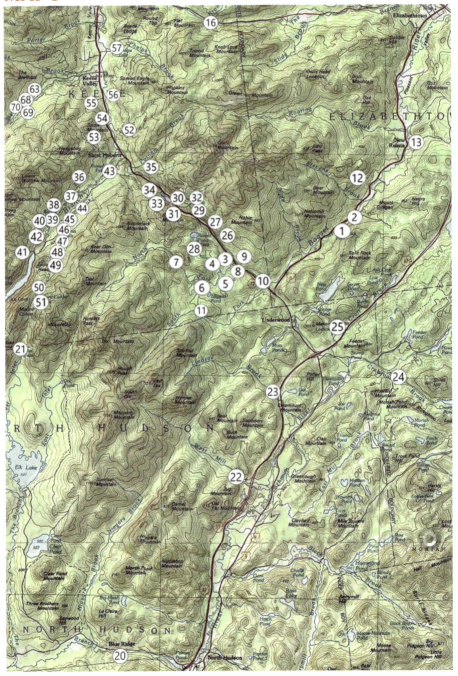

MAP 2

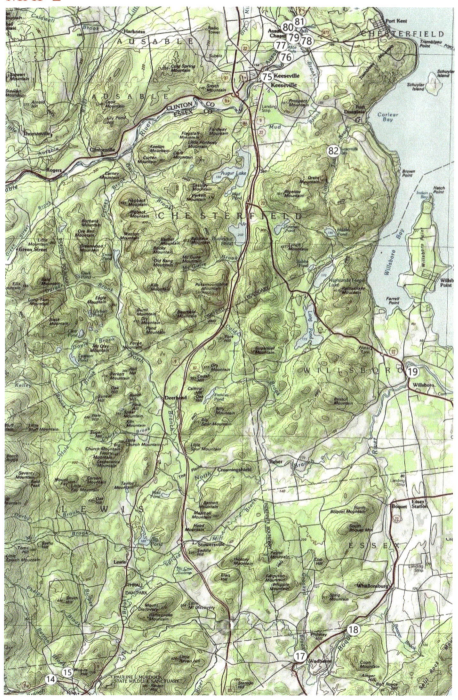

MAP 3

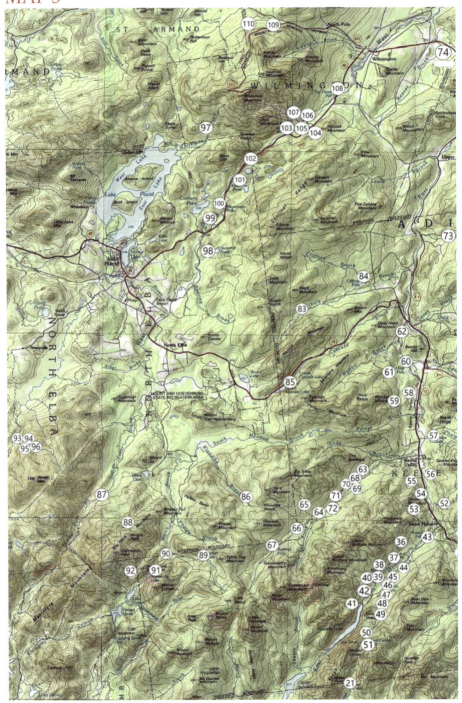

Foreword

Gravity is a remarkable force. I don't know where we'd be without it. More to the point, without it we wouldn't have waterfalls. And then we'd be missing out on one of nature's greatest spectacles.

Think of the Adirondacks as an inverted dome. Here's a project: place a mixing bowl upside down in your kitchen sink. Then turn on the tap and watch what the water does. That's what it does in (or on, or to) the Adirondacks—gravity takes charge, the water runs down all sides, and on all sides it channels itself into waterfalls.

Perhaps we are fascinated by waterfalls because they are so much like us. They are small, and they are big. They are wide and narrow, short and tall. Some are smooth; some have rough edges. No two are exactly alike (I've always thought the ones named Twin Falls must be fraternal, not identical), and yet they are all alike in at least one regard—vertical or practically recumbent, they all transport water downhill. They have unique personalities but congruent dispositions. They act differently in different seasons, hibernating in winter, bursting with new-found vigor in spring, calming down in summer (Russell Dunn has more to say about this in "Why do waterfalls exist?"—which is a very good question). They are creative, shaping pebbles, shaping mountains, shaping Earth if you're patient. What you see and hear is superficial; there is much more to them on the inside once you get to know them.

And they are like us because they can destroy us—they are capable of violence. Like the sirens of mythology, they attract, they beguile, they throw us off guard. They draw us in, and if we are not strong-willed they do not let us out. They must be approached with caution. Be chary of their slippery brinks, the invisible traps in their plunge pools, before you get too intimate.

Or are we so captivated by waterfalls because they are so mysterious? Mountains we can see; most waterfalls, unless they are behemoths like Niagara, lie in wait for us in the deep dense forest, in a gorge, around a bend. Frequently we hear them first; where are they, and what do they look like? We must explore.

And why do we so often speak of them in the plural, when we normally encounter just one of them at a time?

When I was a crewmember at the Adirondack Mountain Club's Johns Brook Lodge during a pair of long-ago summers, I would often spend my weekly day off hiking to a peak or a waterfall. In this way I discovered lovely Beaver Meadow Falls, number forty in this book. I studied it and contemplated how such a simple thing as two molecules of hydrogen plus one of oxygen can spawn such a complex system, beautiful, powerful, perpetual. I strove to focus on one drop of water, maybe splashed off from the main cataract, and thought about the hydrological cycle. I imagined how that drop formed in the skies above and fell onto the Great Range and found its way to those falls, and from there would be carried into the Ausable River, Lake Champlain, the Richelieu and the St. Lawrence, and finally the Atlantic Ocean, where it would evaporate and become part of a cloud again and be carried around the world on the jet stream, ready to tumble onto the Adirondacks once more and perhaps find its way back to Beaver Meadow Falls, like a salmon seeking home. I believe I will return there myself before I die, and see if I can find that drop of water.

Neal Burdick
June 2017

Neal Burdick is the editor of *Adirondac*, the magazine of the Adirondack Mountain Club, and a frequent contributor to regional periodicals such as *Adirondack Explorer* and *Adirondack Life*. He also teaches Advanced Journalism at St. Lawrence University.

Caution: Safety Tips

Nature is inherently wild, unpredictable, and uncompromising. Outdoor recreational activities are by their very nature potentially hazardous and contain risk. All participants in such activities must assume the responsibility for their own actions and safety. No book can replace good judgment. The outdoors is forever changing. The author and the publisher cannot be held responsible for inaccuracies, errors, or omissions, or for any changes in the details of this publication, or for the consequences of any reliance on the information contained herein, or for the safety of people in the outdoors.

Remember: the destination is not the waterfall or the mountain summit. The destination is home, and the goal is getting back there safely.

Aidan Canavan at Rice's Falls. Water is not the only thing that's running.

Waterfall-specific safety tips

1. Always stay back a safe distance from the top of a waterfall. It is natural to be lured to the edge for a better view, but remember that some waterfalls have curved lips at their crest, as opposed to discrete edges; should you start to slide forward, there is nothing to stop you from going over the

brink. Many have died this way. The fact that the bedrock near the top of a waterfall is often wet or slimy further increases the likelihood of losing traction and going over the top.

2. Never jump or dive off ledges near a waterfall no matter how inviting the pool of water below looks. Two things can go wrong: you may lose your footing as you start the jump, causing you to land on the rocks instead of in the water; or you may land on a submerged log or boulder.

3. Be mindful that bedrock can become *very* slippery when wet. Even during dry periods of time, large waterfalls typically send up a spray that dampens the immediate area. Watch your footing and take small steps to increase balance. A hiking pole is always helpful.

4. Never swim in a pool of water located at the top of a cascade. There is always the possibility that the force of the onrushing stream may flush you out from the pool and force you over the top of the fall. Keep in mind that the bedrock is usually worn smooth at the top of a waterfall, leaving you with nothing to grab hold of to save yourself if you get caught up in the current.

5. Never throw rocks over the top of a waterfall. You never know who might be below, even if you just checked a moment ago.

6. Never try to ascend or descend the face or sides of a waterfall unless you are an experienced rock climber or with a climbing guide.

7. Enjoy waterfalls from below whenever possible. Doing so will provide you with the best views, the greatest security, and the most optimum photo shoot.

8. Never drink water from a waterfall no matter how clean the stream might look. *Giardiasis* is too great a risk to assume for the temporary relief of thirst.

9. Be on the lookout for rocks—and blocks of ice during the winter—that could break off and cause serious injury or death if you are standing below. This has occurred in the Adirondacks, and you are particularly vulnerable if you are swimming in the plunge pool of a waterfall. Freak accidents do happen. In 2017 twenty people bathing at the base of Kintampo Falls in Ghana were killed when a massive tree trunk came over the top of the waterfall and crushed them.

10. Always bear in mind that by following a creek up to a waterfall you cannot get lost as long as you follow the stream back down to your starting point.

11. Avoid cornering any wild animal as you enter a gorge to approach a waterfall. You don't want to create a situation where the animal's only path of flight is through you.

Safety tips for all hikers

1. Always hike with two or more companions. That way, if one member becomes hurt, others can stay with the victim while the rest go for help.

2. Make it a practice of taking along a day pack complete with emergency supplies, compass, whistle, flashlight, dry matches, raingear, power bars, extra layers, gorp, duct tape, lots of water (at least twenty-four ounces per person), mosquito repellent, emergency medical kit, sunblock, and a device for removing ticks.

Giants Washbowl as seen from the Giants Nubble.

3. Your skin is the largest organ in your body. To protect it as much as possible, wear sunblock when exposed to sunlight for extended periods of time, especially in the summer, and repellents when you know that you are going into an area brimming with mosquitoes, black flies, and other biting insects. A substitute for applying repellents is to take a lead from horses and pack a leafy branch that you can wave around you to keep the annoying insects at bay.

Wearing a hat with a wide brim to keep out the sun is always helpful, too. Remember that you can get burned even on a cloudy day. Wear long pants and a long-sleeve shirt to protect yourself from both the sun and biting insects.

4. Wear ankle-high boots—always! Boots provide traction, gripping power, and ankle support that your sneakers and shoes cannot provide. People are especially asking for trouble when they enter the woods wearing loafers or flip-flops,

To reduce the likelihood of blisters on long hikes, wear liners under wool socks.

Beaver Meadow Falls. Photograph by John Haywood.

5. Be aware of the risk of hypothermia, and stay dry. The air temperature doesn't have to be near freezing for you to become over-chilled.

Equally of concern is the danger of hyperthermia (overheating). Be sure to drink plenty of water when the weather is hot and muggy. Also, stay in the shade whenever possible, and use the stream near a waterfall to cool off in if you begin to overheat.

6. It is best to stay out of the woods during hunting season, but if you do enter the woods during those times, wear an orange-colored vest and make periodic loud noises to draw attention to the fact that you are a human and not an animal.

7. Stay on trails whenever possible to avoid the chance of becoming disoriented and lost. Off-trail hiking also causes more damage to the

environment, particularly where there are rare plants and mosses off-trail as there are on some of the Adirondack High Peaks. Do not consider bushwhacking through the woods unless you are an experienced hiker, you have a compass or GPS unit (including spare batteries) and know how to use them, you are prepared to spend several days in the woods if forced to by circumstances, and you are with a group of similarly prepared hikers.

Lower Ausable Lake from Indian Head.

8. Be flexible and adaptive to a wilderness environment that can change abruptly. Trails described in this book can become altered by blowdown, beaver ponds, avalanches, mudslides, and forest fires.

9. Always let someone know where you are going, when you will return, and what to do if you have not shown up by the designated time.

10. Avoid creatures that are acting erratically. Any animal advancing toward you that cannot be frightened off must be assumed to be either rabid or operating in a predatory mode.

11. Use good judgment—something that is sometimes easier said than done. Unless it is critical that you react immediately and decisively, if you encounter a problem stop for a moment and think through what your options are. This is never truer than if you suddenly find yourself lost or disoriented.

The old sports adage "the best offense is a good defense" applies to hiking. It's far better to defuse a problem early on than to wait until it has reached crisis proportions.

12. Leave early in the morning if you are undertaking a long hike. You don't want to be caught out in the woods with daylight dwindling and a long ways still to go. Allow for even more extra time if the hike is in the winter, where night arrives hours earlier than it does in the summer.

13. Be mindful of ticks, which are becoming more prevalent and virulent as their range increases. Check yourself thoroughly after every hike and remove any tick immediately. The longer the tick remains in contact with you, the greater the risk that you will contract a disease such as Lyme.

Special precautions for children and pets

This subject takes us back to the underlying premise that waterfalls must always be respected and viewed as wild unbridled natural wonders. Waterfalls can kill.

No one really knows for sure how many people have died at Niagara Falls over the last three centuries, for many deaths have occurred without witnesses present. We do know, however, that by the end of the nineteenth century Niagara Falls had not only become the honeymoon capital of the world but its suicide capital as well. (This dubious distinction only ended when the Golden Gate Bridge in San Francisco, California, was constructed in 1937.) Big waterfalls—for example T-Lake Falls near Piseco Lake—all have deaths associated with them, generally because a certain percentage of visitors inevitably engage in horseplay, try reckless stunts, get too close to the top of the waterfall, or choose to tempt fate by rock climbing where they shouldn't.

If you decide to approach a large waterfall near its top, keep young children carefully supervised and dogs leashed at all times.

Mileage

You should find that the mileage indicators given for reaching the trailhead are reasonably accurate and that the use of clearly identified intersections to start from will eliminate any ambiguity about whether or not you're taking the right route. The mileages stated are generally accurate to within +/– 0.1 mile, but please bear in mind that there can be differences in readings between individual odometers.

Mileage indicators given along hiking trails are less exact and frequently estimated, and therefore they may not be as accurate as the distances measured by odometer. That said, care was taken to ensure that they are accurate enough to tell you where you are along the trail and how to reach your destination. Hiking mileages are one-way only unless otherwise specified.

All distances given are in feet, yards, or miles, even though most of the rest of the world has converted or is ineluctably moving toward conversion to the metric system. For visitors to the U.S.A., the following conversions may be helpful: 1 foot = 0.3048 meter, 1 yard = 0.9144 meter, and 1 mile = 1,609 meters (or 1.6 kilometers).

Some nineteenth-century writers gave distances in terms of rods. The conversion is: 1 rod = 16.5 feet.

You will find that GPS coordinates for parking and destinations are liberally distributed throughout the book. Most were taken right on the spot using a handheld GPS unit. To confirm their accuracy, however, coordinates were double-checked using Google Earth and adjustments made where necessary.

Introduction

I am the author of seven regional waterfall guidebooks (with more still to come), but the area of the Adirondacks that this book covers holds special meaning to me, as it does to my wife and fellow author, Barbara Delaney. For the last fifteen years we have been leading a weekend exploration of early-spring waterfall-related hikes in the Greater Keene Valley area from St. Huberts to Lake Placid (considered by some to be the "Switzerland of America") and from North Hudson to Keeseville. We call it "Waterfall Weekend."

The information on the majority of waterfalls presented in this book has come from those years of exploring the region and then putting together a series of annual hikes to new and different waterfalls. Waterfalls chosen for each outing had to be within a reasonable driving distance from Keene Valley, which served as our starting point, and not so arduous a hike as to dissuade our guests from following behind us. As a result virtually all of the destinations described herein involve day hikes and are well within the reach of accessibility for your average half-day hiker.

"Why Keene Valley?" you might ask. The obvious answer is that Keene Valley, nestled in the Highs Peaks region of the Adirondacks, is home to "The Garden," one of three main entry points into the High Peaks and its waterfalls. We also simply fell in love with Trails End Inn, which became our home base, and its innkeepers, Sue Lindteigen and Dave Griffiths. Not once did we ever consider abandoning Trails End Inn, a converted 1902 lodge in Keene Valley, for another inn at another location in order to increase the number of potential waterfalls within a reasonable driving distance. Our loyalty remained unwavering year after year.

What makes the Greater Keene Valley area so appealing to waterfall lovers are its countless waterfalls—a fact that may not be apparent at first glance. You can easily drive through the region and see only Roaring Brook Falls in St. Hubert, a roadside cascade south of Keene Valley, Cascade Lake Falls at the Cascade Lakes (if you're lucky), High Falls Gorge (a commercial attraction), The Flume northeast of Lake Placid, and Rainbow Falls at Ausable Chasm. But there is so much more.

Consider this book, then, as a journey wherein we take one small slice of the Adirondacks—the Greater Keene Valley /Lake Placid/ Keeseville region—

and magnify it into a huge roadmap for hikes of discovery to find waterfall after waterfall after waterfall.

Why write about waterfalls?

Why did Barbara and I become interested in searching for and writing about waterfalls? Like so many others in pursuit of adventure, we started off with a normal amount of curiosity and interest. What tipped the scales for us was not just one "eureka!" moment, but rather a slow, deepening evolution.

Fairy Ladder Falls. Note the hiker standing next to the waterfall. Photograph by John Holmes.

At first out main interest was in caving (but not as hard-core cavers). We ventured out to explore underground wonders in Albany County, Schoharie County, and the Berkshires. After Barbara had a bad experience in Single X Cave in Schoharie County, this phase of our lives came to an end.

At the same time we were caving, we were also hiking up mountains in the Adirondacks, Catskills, Shawangunks, and Berkshires. This was easy enough to do since we live in Albany, at the hub of the wheel. In the end I suppose that hiking to waterfalls was an inevitable attempt to find the midpoint between the lows of caving and the highs of mountain climbing, and waterfalls slowly became our main interest.

It also became fun to try to find obscure waterfalls in eastern New York State that had vanished from contemporary awareness. For us, seeking out these forgotten waterfalls was equivalent to going on a treasure hunt. We began to scour old topographical maps and state gazetteers; we collected antique postcards that featured long-forgotten waterfalls by name, and we pored over old-time stereoviews; we talked to local town historians and to knowledgeable elders; we went on Google Earth to visually hunt for waterfalls; we visited the New York State Library to rummage through old books for passing references to waterfalls; and we even talked with many local youths, who were frequently better acquainted with natural areas of interest than their elders.

Vortex on the East Branch of the Ausable River. Photograph by John Haywood.

Through this process we found many waterfalls that had fallen off the grid, long-forgotten by both locals and tourists and known only to a few backwoodsmen. Of course, we have also had our share of disappointments. We have hiked into many ravines, gorges, and along streams where topographical maps had suggested the possibility of a waterfall, only to find nothing. As the reader you are spared all of these dead ends, although we will tell you about several of the adventures we have had when we came back empty-handed.

Today there is growing concern that the Adirondacks are being loved to death, with hordes of hikers descending on the Keene Valley area every weekend and sometimes even in large numbers during the week. In early

September 2016, sixty-seven Canadian hikers arrived in two buses and proceeded to climb up to the summit of Algonquin Peak. The huge party was reported to the authorities by other hikers, resulting in a ticket being issued to the group leader and organizer, a thirty-four-year-old Quebec man, for bringing in a group that exceeded fifteen hikers. Too many hikers congregating in one place spoil the wilderness experience for all.

Many hikers entering the High Peaks region are driven to reach the summits of towering mountains for the "big picture" views. I, too, have felt the powerful allure of mountains. It is my hope, however, that a book devoted exclusively to waterfalls will offer exciting alternates, ones where fewer people are likely to gather than on the summit trails.

Readers should be aware that several chapters describe waterfalls that are located on private property. They have been included both for the sake of thoroughness and for their historical significance. "No trespassing" signs should always be respected, and the land given wide berth unless permission to enter has been obtained from the landowner.

A few of the chapters involve bushwhacks. However, these bushwhacks don't involve tramping through featureless woods with map and compass (or GPS unit) in hand, or undertaking what Orson Schofield Phelps, the famous nineteenth-century Adirondack guide, would have called "random scoots" (trail-less) as opposed to "reg'lar walks" (on trails). Because waterfalls are located on streams, you need only follow the stream up to a waterfall and then back down again. While it is true that you must still bushwhack through thickets, blowdown, and uneven terrain without a trail, it is unlikely that you will get lost as long as you stay in close proximity to the stream, your lifeline both in and out.

Monument Falls along Route 86.

I am constantly reminded of the famous quote attributed to Sir Isaac Newton, "If I have seen farther, it is only by standing on the shoulders of giants," and I have tried to live by it accordingly. All too frequently, guidebook writers neglect to credit the many authors, photographers, and explorers who have gone before them. Very few of us are likely to cover a subject that hasn't already been touched upon by others at some earlier point in time. For this reason I have made every effort to give credit where due and to include descriptions of waterfalls from past authors, including references to photographs dating as far back as the 1800s.

A number of interesting articles have been written about Adirondack waterfalls over the last quarter-century. One of the first was Clyde H. Smith's "Adirondack Waterfalls" in the spring 1974 issue of *Adirondack Life*. Smith valiantly attempted to list and categorize many of the Adirondacks' waterfalls using a roster compiled by Greenleaf Chase, breaking them down according to ownership, county, town, and 15' quad. This list, which was later reprinted in Howard Kirschenbaum's 1983 book *The Adirondack Guide*, barely cracks the surface, but it's a first try and a good one for sure. Included is a wonderful colored picture of Beaver Meadow Falls complete with a hiker standing near the base to give the reader a sense of the size and dimension of the waterfall. Also included are colored photographs of Roaring Brook Falls (which Smith tentatively states has "a total drop of more than 100 feet") as well as O.K. Slip Falls (which, like Hanging Spear Falls, Lampson Falls, Salmon Falls, and many others is outside the purview of this book).

In the March/April 1981 issue of *Adirondack Life*, an article by Paul Schaefer titled "A Sound of Falling Water" discussed thirty major rivers ringing the Adirondack Park, each containing its share of waterfalls. Schaefer concluded by offering a very wise tip about when best to visit waterfalls if you want to see them at their mightiest—in April, May, and early June, or in September following rainfall.

In the *Adirondack Life 1999 Annual Guide to the Adirondacks* an article called "Liquid Assets" presented a variety of waterfall and flume photographs, including a partial shot of Wanika Falls by Jon Nedele and a photograph of Chasm Falls on the Boquet River by Frank Houck.

An article by Barbara McMartin titled "The Other Niagara Falls: Tough hikes to hidden cascades" appeared in the *Adirondack Life 1999 Annual Guide to the Adirondacks*. Two major bushwhacks to waterfalls are described, one being on Niagara Brook and the other on a stream flowing off Macomb Mountain.

These are the articles written about Adirondack waterfalls prior to this millennium that come most readily to mind. No doubt there are other articles that I have overlooked but that would also deserve recognition. Clearly, people have been thinking and writing about Adirondack waterfalls and photographing them for a long, long time.

At some point it's entirely possible that you may want to strike off on your own to explore the great outdoors and go looking for waterfalls. I have given this matter some serious thought over the years and wrote about it in the spring 1997 issue of *Kaatskill Life* magazine in an article called "Hunting for Waterfalls," wherein I offered tips on how to go about finding waterfalls on your own.

Waterfalls will always hold special meaning for Barbara and me. We are still enthralled every time we come across a waterfall whose existence was previously unknown (to us). We hope that this book will help you to share this excitement with us.

Why do waterfalls exist?

If the world were as perfectly round as a billiard ball and not continuously uplifted by plate tectonics, then there would be no mountains or hills, only a planet-girdling ocean—and certainly no waterfalls. If the Earth didn't rotate, producing uneven temperatures on its surface, then there would be no winds to carry moisture-laden clouds over land masses—and no inland rain to produce waterfalls. If the planet were made of uniform bedrock without sections of durable rock mixed with easily erodible rock, then streams would run down mountains creating unvarying riverbeds without drops or plunges—and there would be no waterfalls.

View from former Marcy Dam.

If our DNA reproduced perfectly, then there would be no random mutations—hence, no evolution and no you or me to appreciate the beauty of waterfalls.

We can all be thankful, then, that the Earth is far from perfect. It is through Earth's imperfections that waterfalls of extraordinary beauty are produced and that we are here to respond to them as we do.

Just like drifting snowflakes, no two waterfalls are ever exactly alike. Some are towering cataracts; some are tiny and seemingly of little consequence. Some are shaped like staircases with many ledges and drops; some glide down through extended chutes, producing waterslides. Some plunge off the tops of escarpments into deep pools; others drop onto blocks of talus. Some are miles wide (yes, there are actually waterfalls of such dimensions), and some are narrow and can be easily crossed with a single stride. Some are found in deep gorges, while others are formed on streams that barely scratch the Earth's surface. Some are brimming with water most of the year, while others are a mere trickle except during spring's run-off.

Even the same waterfall never remains the same from moment to moment. The Greek philosopher Heraclitus said, "You cannot step twice into the same river"; the river has already changed by the time your foot reenters it. So it is with waterfalls. In the spring, waterfalls are loud and boisterous, awakening from their slumber and bursting forth to announce their presence.

Rock Garden Falls. Photograph by John Haywood.

Some are so powerful that their waters hit the ground like pile drivers. This is the season when waterfalls are at their mightiest. In summer, waterfalls turn languid, wrapped up in the greenery of life as it bursts forth. They are as much entities of rock as of water in the warmest season. Come autumn, waterfalls reanimate as trees pirate away less groundwater for wood production. Autumn brings vivid colors and a lack of pesky biting insects. In winter, waterfalls go into a deep sleep and hibernate, encased in sheets of blue-green ice. The only sound likely to be heard in the cold, crisp air is the faint gurgling of water flowing under the ice pack, or the whack of an ice-climber's hand ax as they make their way up the tower of frozen water.

And then comes spring, and the whole cycle begins again.

The Adirondacks

"Real is what Earth is. Real is what people are. Unreal is the separation of the two."
— Phil Gallos, Adirondack author, 1972

The High Peaks Region represents an enormous drainage system containing thousands of miles of streams and rivers that conduct water toward the Hudson River to the south, the St. Lawrence River to the north, the Great Lakes to the west, and Lake Champlain to the east. With such an abundance of flowing water, it is no wonder that the Adirondacks have produced waterfalls of such sublimity and magnificence.

Native Americans foraged and hunted in the Adirondacks, but they never took up permanent residence there. Two regional waterfalls are named Indian Falls, but there is nothing in regional history to suggest that there actually was a Native American presence at these sites.

Europeans were the first permanent settlers. They initially came to farm the lands. The harsh weather—particularly the winters—turned most of them to hunting and trapping for a living. It's doubtful that they spent much time thinking about waterfalls. More likely, just the necessities of survival occupied much of their waking time.

In 1837 Ebenezer Emmons published a geological survey that demolished the myth once and for all that the Catskill Mountains were higher than the Adirondacks. In fact, the highest mountains in the Adirondacks turned out to be not only higher, but nearly a thousand feet higher than the highest peaks in the Catskills. This discovery undoubtedly spurred greater interest in the Adirondacks and, it stands to reason, in its waterfalls as well.

In the mid-1800s some of the early settlers turned to part-time guiding as a way to supplement their income, taking advantage of the influx of visitors (called "sports") who came in droves after being inspired by the writings of William Henry Harrison Murray, Seneca Ray Stoddard, J.T. Headley, Charles Fenno Hoffman, and other nineteenth-century authors. According to Roderick Nash in *Wilderness and the American Mind* (2001), "By the 1880s, more had been written about the Adirondacks than any other wilderness area in the United States."

Tourism was also stimulated by the many artists who came north to render onto canvas the incredible scenery of the Adirondacks—artists such as Asher B. Durand, John Casilear, John F. Kensett, Frederick Perkins, Roswell Shurtleff, A.H. Wyant, Winslow Homer, Sanford Gifford, John Fitch, John Adams Parker, Samuel Coleman, Jervis McEntee, and Arthur Parton. Since waterfalls were often featured in these artworks, they became a vital piece of the Adirondacks' lure as well.

Around the same time or earlier, mining and logging became major Adirondack industries. Lumbermen established sawmills next to waterfalls and built dams to generate hydropower. Forges were built to melt iron ore into malleable metal. Waterfalls not only were part of the Industrial Revolution, they drove it. The full impact of this realization led me to write an article for the 2005 issue of the *Catskill Mountain Region Guide* called "The Power of Waterfalls" and for the April 2010 issue of the *New York State Conservationist* called "Workhorses of the Industrial Revolution." It is important to remember that prior to the discovery of fossil fuels there were few dependable energy alternatives other than waterpower.

Wedge Brook Falls. Photograph by John Haywood.

Lumbermen built hundreds of miles of logging roads into the interior of the woods in order to harvest the forests. We have benefited greatly from those early roads, for today many of them still serve as pathways along streams and to waterfalls.

From the late nineteenth century to the early twentieth century, postcards and stereoviews were reproduced by the millions, popularizing many natural sites including waterfalls. Stereoviews served as a form of Victorian television, transporting viewers to areas of beauty as they sat comfortably in their armchairs. Certainly, many people were subsequently inspired to go out and visit waterfalls.

Today many of our waterfalls are at the best they have looked for the past three centuries. Gone are the unsightly mills with their sluiceways, pipes, bricks and mortar, stricken down either by the willful hand of humans or by Nature herself. It is an ideal time to get out and explore what this region has to offer.

Mister Indestructible?

I am often asked if I have had any close calls during my years of searching for waterfalls. After all, many of the hikes I undertake are solo endeavors and sometimes lead to places far from roads and paths. Many involve descents into steep gorges where waterfalls may or may not lurk. Some require scampering up streambeds filled with stones and giant rocks. Nevertheless, I must answer with a carefree shrug that I am basically a pretty dull fellow. I really never have had any close calls.

Roadside Falls. Photograph by John Haywood.

That's not to say that things haven't happened. The incident that most frightened me occurred when I was in my early thirties and heavily involved in caving. It was a late summer afternoon and I was dressed in shorts and a T-shirt. I had gone to Clarksville, a small village near Albany, to look for the Clarksville Cave. My goal was simply to locate it and then revisit the cave at a later date with my caving gear. I found the cave readily enough. Its entrance was a hole in a large sinkhole. Curiosity got the better of me. I had a flashlight with me, so I thought, "Why not go down, take a look around the chamber that the wormhole leads into, and then pop back up to the surface. Nothing could possibly go wrong."

So I lowered myself into a fairly tight, descending passageway and, sure enough, immediately came out into a large chamber. Using the flashlight to illuminate the interior, I looked around for a moment or two. "Cool!" I thought. Having seen what I wanted to see, I started back up the passageway and immediately ... hit bedrock! Unexpectedly, there was solid rock above me. The passageway up to the surface was no longer there! Suddenly, my predicament became all too real. Here I was, thirty feet underground in a fifty-degree chamber wearing shorts and a T-shirt. Nobody on Earth knew where I was. The only thing standing between me and never getting out of this cave was the flashlight in my hand, whose light suddenly seemed to be dimming. I began to hyperventilate as panic swept over me like an icy stream of cold air. I hurried back down to the bottom of the chamber and once again started up, this time heading off in a slightly different direction. After a terrifying moment the passageway kept going and I found myself back up on the surface, literally ready to kiss the ground.

That's been my most harrowing experience.

I have never gotten seriously hurt. Ever. Although, like everyone who has hiked a lot, I have taken my share of tumbles and falls. The closest I ever came to getting seriously injured was in Austin Glen, near the village of Catskill. I was scampering down a rocky slope to reach the riverbed. For reasons I can't explain, I was directly facing forward toward the river as I made my way down, instead of stepping sideways so that if I fell I would just slide and not tumble. My right foot stepped onto something that gave way, and I found myself pitching forward headfirst onto the rocks, five or six feet below. I hit the rocks and boulders with as much grace as a tree crashing onto the ground. Perhaps I instinctively twisted in such a way as to minimize the impact. It all happened so fast that I can't remember the details that seemingly took place in a micro-second. All I know is that my whole body was suddenly wracked with pain. I was convinced that I must have broken something. As a general rule I tend to jump right back up onto my feet after taking a fall, believing (or self-deceiving) that I can trick my body into thinking that nothing untoward happened. This time, however, I lay there on the rocks for several minutes while taking inventory of all my parts. Gradually I realized that, while I may be bruised and battered, nothing had broken. I got up, shook myself off, ignored the pain as best as I could, and managed to get myself out of the gorge the way I had come in. Within a week or two the bruises and scrapes were gone, but a puncture wound on my back lingered for months. I still cringe when I think about how lucky I was that I didn't hit my head. The concussion would have knocked me out for sure—or worse.

People who hike with me generally come away unscathed. One incident almost turned tragic, however. I was hiking with my good friend Christy

Butler while doing field research for a book on glacial boulders of Massachusetts that we were writing together, called *Rockachusetts: An Explorer's Guide to Amazing Boulders of Massachusetts.* At the time, we were snowshoeing along a cliff edge on a multilevel escarpment looking for two natural wonders called The Pillar and Sunderland Cave. I had gone ahead and had just rounded a corner when I heard Christy cry out. I rushed back quickly, fearing the worst. Christy was no longer on the trail. He was clinging

Cascade Lake Falls in the early morning mist. Photograph by John Haywood.

to a shrub on a nearly vertical slope about fifteen feet below where I was standing. Somehow his snowshoe crampons hadn't held, and he slid right off of the trail and over the edge of the escarpment. To make a long story short, we got Christy back up to the ledge again and, other than being shaken up a bit, he was none for the worse. A close call.

See, I really don't have any life and death stories to tell. And you know what? As far as I'm concerned, I'm perfectly happy to let it stay that way.

John Haywood's Tips on Photographing Waterfalls

Everyone loves waterfalls. People love the scenery, the sound, and the feeling they get as the water crashes about the rocky landscape whipping up a misty breeze. Here are some pointers on how to capture all of that in a photograph.

Photographing waterfalls is no different than photographing a landscape. All you need are the proper tools, and you're all set. That being said, do not feel you need the latest and greatest, most expensive, camera and lens. A great photograph can be made with any camera.

The tools:
- Camera and lens
- A sturdy tripod (a bean bag or something similar will also do the trick)
- Neutral density filter—a dark filter that allows for slower shutter speeds
- Polarizer filter—a filter that removes glare from water and wet rocks and that will also darken by about one f-stop
- Remote shutter release—a wired or wireless controller that allows hands-free shutter release of the camera in order to decrease camera shake (if you do not have one, you can use the timer on your camera)

Setting up: When you reach your destination, take a few minutes to scope out the terrain. Notice the lighting, the shadows, patterns in the trees, and any interesting formations in the rock or flow of water. I like to take a moment to "feel" the scene. Taking a few deep breaths and just absorbing all that surrounds you can help you concentrate and focus.

When you have found your shot, set up your tripod and camera, attach whichever filter you prefer to use, and set your camera to "Shutter Priority." This setting will allow you to choose the shutter speed you want while the camera adjusts the aperture. Also, set the ISO, which is the light sensitivity, to 100 or whatever the lowest value is on your camera. This will allow for use of a slower shutter speed. Most important—do not just focus only on the waterfall. When I started photographing waterfalls, I would zoom in on the waterfall and leave out all of the beautiful surrounding scenery. Remember, you want to tell a story.

Making the photograph: Look through the eyepiece and frame your shot. Be sure to look everywhere in the viewfinder for any details you may want to include or crop out by moving the camera or zooming in or out. Now decide on the look you want the water to have. What the human eye sees is between 1/60 and 1/30 of a second. This will freeze the water with minimal movement (Image A). A much faster shutter speed will virtually freeze the water in place with zero movement. A slower shutter speed will show much more motion. A lot of people like to create the "smoky" or "creamy" look of the water by using the slowest shutter speed (Image B) they can get, while others prefer the more natural look. It all depends on *your* vision of what you want *your* photograph to look like. I have found that a shutter speed between 1/6 and 1/8 allows for a nice blending and shows the motion of the water while also allowing some of the smooth effects of a slow shutter speed (Image C). Again, this is entirely up to you as the photographer. The nice thing that digital cameras offer is the ability to see your photos and make adjustments right then and there. Don't be afraid to take advantage of this feature; experiment by using different settings. Also, don't be afraid to make some mistakes when experimenting. Purposely set the exposure too high or too low to see the results. Try different angles, and always remember also to make some photos in the vertical.

Photography is not as difficult as many think. If you have the eye for a good shot and the patience to learn, the world can be your canvas.

About John Haywood: I have known John Haywood for a number of years and have collaborated with him on eleven different projects, some involving stereography. Haywood is a highly regarded photographer best known for his Adirondack waterfall calendars and his Web site, jhaywoodphoto.com. He is also the Executive Director of the New York Waterfall Conservancy, an organization that is dedicated to the preservation and beautification of New York State's waterfalls. Haywood is a very creative individual, his mind a ceaseless fountain of new ideas. It is my good fortune that he has graciously contributed a number of his waterfall photographs to this book as well as providing the front cover shot.

Russell Dunn

Cascade Lake Falls. Photograph by John Haywood.

I: THE BOQUET

Split Rock Falls. Photograph by John Haywood.

The Boquet is a 40-mile-long river with a 280-square-mile watershed. It rises in the High Peaks Region of the Adirondacks, principally from the upper reaches of Dix Mountain, and drops nearly 3,000 feet as it rushes madly downhill into the valley. Because it is the steepest river in New York State, the Boquet produces some dynamic waterfalls along its descent. Its steepness is also the reason why the river is prone to flash floods and heavy erosion.

In 1982 the decision was made by the Board of Geographic Names to fix the spelling of the river to "Boquet," eliminating an earlier spelling of "Bouquet." Boquet is pronounced "Bo-Kwet," although some still continue to pronounce the name as you would a bouquet of flowers.

There is considerable dispute as to the origin of the river's name. Some contend that it arose from the river's trough-like mouth, which the French call *baquet*; others, from the flowers lining the river's bank. Some insist that it came from General Boquet, a British officer in the French & Indian War (but one who never visited the Champlain Valley); others, from Charles Boquet, a French Jesuit novitiate whose mission was to convert the Iroquois to Christianity. From my research the first explanation seems to be the one most favored by scholars.

You can follow the Boquet River by car, starting from where Route 73 crosses the river's North Fork all the way down to where the Boquet flows into Lake Champlain at Willsboro. Along the way you will come across a number of vantage points where you can get out and walk a short distance to see a Boquet River waterfall.

The South Fork of the Boquet (which rise up by South Dix and East Dix) and the North Fork of the Boquet (which rises northwest of the Rock of Gibraltar) come together just northwest of the junction of Routes 9 & 73 at what has been playfully called Malfunction Junction because of its maze-like crisscrossing of roads. If you know what to look for in advance, the confluence of these two branches can be observed to your right as you head northwest on Route 73. The North Fork produces a number of waterfalls that are accessible by trails. This is not the case with the South Fork.

It is a point of historical interest that a small battle was fought at the terminus of the Boquet at Willsboro in 1814 during the War of 1812. Here, the British were driven off by a highly motivated and quickly assembled band of Essex County militia.

The main organization looking after the health and well-being of the Boquet today is the Boquet River Association (BRASS), whose base of operation is in Elizabethtown. The association was formed in 1984. Their Web site is boquetriver.org.

1. SPLIT ROCK FALLS

"There is a hidden message in every waterfall. It says, if you are flexible, falling will not hurt you!"—Mehmet Murat ildan (Turkish playwright and novelist)

Priority #1—Staying safe: We start at Split Rock Falls, midway along the Boquet River and arguably the Adirondacks' most notoriously dangerous waterfall. But just what led to its infamy?

To better understand some of the realities, Split Rock Falls is the largest accessible roadside waterfall in the Adirondacks, complete with its own parking area. It is also one of the most popular and heavily visited non-commercialized recreational spots in the Adirondacks. On a hot weekend day in the summer, the waterfall is crowded with sun-worshippers. With greater numbers of visitors comes greater potential for tragedy.

What makes the waterfall potentially dangerous is that it is blessed with high ledges and deep pools, just the combination of features that you would expect would encourage revelers (usually teenagers and young male adults) to jump from its heights. Not surprisingly, this leads to broken bones and, on occasion, death. There is a small rock outcrop with a fifty-foot drop into a pool that locals call "Suicide Leap."

Remember—safety first.

Whitewater enthusiasts even occasionally paddle over Split Rock Falls. A photo of this can be seen in Don Morris's article "Creek Freaks" in the June 2007 issue of *Adirondack Life*. Paddlers consider the waterfall to be Class V–VI whitewater.

But it is for one terrible incident, and that one terrible incident alone, that Split Rock Falls achieved its notoriety. In 2003 a group of camp counselors from Camp Baco in Minerva came to see the falls. Coincidentally, an incredible deluge of rainfall had swelled the Boquet River just before their visit, turning the normally sedate falls into a roaring colossus. Visitors later stated that the falls sounded like a freight train thundering through the notch. None of the Camp Baco counselors had planned to go in swimming—that is, until David Altschuler (age eighteen), one of the youths and a very good swimmer, lost his footing and tumbled into the maw of what looked like a giant washing machine. Altschuler immediately was in serious trouble and instantly vanished from sight. Without forethought or hesitation, his three friends—Jonah Richman (eighteen), Jordan Satin (nineteen), and Alan Cohen (nineteen), excellent swimmers all—jumped in to rescue him.

In the end, not only couldn't they save David Altschuler—they couldn't save themselves. All four perished.

What had happened?

A vortex of incredibly tragic circumstances had created a deadly trap at the falls. Most of us, except for whitewater paddlers, are unaware of just how much force moving water can exert as its speed and volume increases. On that particular day the volume of water pouring over Split Rock Falls was torrential, exerting a force greater than 2,000 pounds per square inch. Against such an onslaught, even an Olympic weightlifter cannot push his way out.

But that wasn't all. The water was so churned up that the surface had become the consistency of froth—more bubbles and air than fluid. There was nothing that the flailing arms of a swimmer could push or pull against to get out.

Perhaps also, like Huntington's Gorge—Vermont's most notorious chasm that has claimed the lives of over twenty-five victims—unseen ledges and overhangs may lurk below the water's surface where a victim, once pushed down, ends up pinned and unable to fight his way back out.

All these factors contributed to one degree or another to what was an inescapable deathtrap. Particularly sad is that what was a tragedy for one youth instantly escalated into a catastrophe that claimed three more lives.

It's sobering to realize that a calamity of this magnitude can happen to anyone, anywhere, at any time. Imagine, just for the sake of argument, that you were one of the bystanders. What would you have done? Would you have stood there passively, watching helplessly as another human being drowned? Faced with conditions that transcend our normal experiences, it is hard to say what we would do. Many of us would be tempted to jump in to help or to get dangerously close to the maelstrom in order to lend a hand (and thus possibly get pulled in). How can we know for sure when conditions are so extreme that they are beyond our survivability? This is a question that I leave for each reader to answer for themselves.

When you visit Split Rock Falls today, the odds are 10,000 to 1 that the waterfall will be anything other than its old friendly self—not a raging, foaming killer. But waterfalls deserve respect, and sometimes it turns into a matter of life and death when they are given less than their just due.

Split Rock Falls features three drops and two pools. Photograph by John Haywood.

This is a matter to which I have given a considerable amount of thought, ultimately resulting in a piece I wrote for the spring 2005 issue of *Kaatskill Life* entitled "Safety First at Waterfalls." It should be mandatory reading for all who love waterfalls and seek to visit them.

Enjoying the falls: Split Rock Falls was named for the manner by which the bedrock splits the river momentarily into two channels, causing it to rush over three cascades and into two pools. The total height is no more than thirty feet. The falls are tucked away in a deep valley created by the Boquet River, with Split Rock Mtn. (~1,950') to the southeast, and Holcomb Mtn. (~2,303') and Deer Mtn. (~2,232') to the northwest.

In his 1874 book, *The Adirondacks Illustrated*, Seneca Ray Stoddard describes the falls as being "where the water comes sparkling and foaming down through a gorge and over the rocks, descending about a hundred feet."

(It would seem from the description given that Stoddard is also including the lower fall, 0.2 mile farther downstream).

A twenty-first-century description of the falls is found in Hamilton Davis's article, "A River Rebounds," in the June 2003 issue of *Adirondack Life*, where the author writes, "The river swirls over and between the fractured rock, then plunges wildly down the hill in a staircase cascade, diving fifteen feet into a plunge pool, gathering itself briefly then crashing into a second basin, then a third before splashing into the wide holding pool on the valley floor."

Native Americans were the first people to visit the area. They believed that the unhappy spirit of a drowned chief dwelled by the falls. In order to placate the spirit, they pushed leaves filled with tobacco (a fairly precious commodity in those early days) onto the waters before setting off in canoes farther downstream. Undoubtedly, this worked very well since nothing was likely to happen anyhow.

Like many waterfalls during the eighteenth and nineteenth centuries, Split Rock Falls was used for hydropower. In 1825 (the same year as the grand opening of the Erie Canal), Basil Bishop erected a cold-blast forge at the falls. An old dam can still be seen at the top of Split Rock Falls.

New York State acquired the falls from Richard W. Lawrence, Jr. in the 1980s. To promote safety, the state erected a chain-link fence around the falls' most dangerous areas, but the fences have not stood the test of time as barriers, for partiers have broken through them over the intervening years. Their presence today is more an unsightly nuisance than a deterrent or safety guard.

As a point of historical interest, an inn south of New Russia once capitalized on the waterfall's name. In the 1930s, Otto de Muth and his wife purchased a large home that had earlier been called Beaver Meadow Farm and then Old Tavern House. They changed the name to Split Rock Inn, which endured as a business into the 1950s.

Photographs of the waterfall can be seen in Den Linnehan's *Adirondack Splendor* (2004), Randi Minetor's *Hiking Waterfalls in New York* (2014), and Mark Bowie's *The Adirondacks: In celebration of the seasons*.

To get there: From Underwood (junction of Routes 73 & 9), proceed northeast on Route 9 for 2.3 miles. As soon as you cross over a bridge spanning the Boquet River, turn right into a large parking area (44°07.453'N 73°39.475'W). Be mindful of the sign that says "30 minute parking," an obvious attempt to limit people from coming to the waterfall and partying all day. Remember, alcohol and water on the rocks is one cocktail that can be a lethal mix.

0.2 mile farther down the road is a curious roadside memorial on your right. A sign by an unnamed rock monument states, "A life fully lived shines on."

2. UNDERWOOD FALLS

Less than 0.2 mile downstream from Split Rock Falls is a second large waterfall. It is increasingly being called Underwood Falls based upon its proximity to the hamlet of Underwood, but it could just as easily be called Lower Split Rock Falls. Underwood Falls rivals Split Rock Falls both in terms of size and power, but has one additional note of distinction. It is located in a

Underwood Falls, aka Lower Split Rock Falls, is less than 0.2 mile downriver from Split Rock Falls. Photograph by John Haywood.

deep gorge with high walls, making access more problematic. The east wall of the gorge is nearly vertical; the west wall is less so. On entering the gorge, the Boquet, previously flowing along languidly after leaving Split Rock Falls, accelerates as the walls press in, producing a powerful cascade. This is not a gorge to be caught in during times of high water or in the early springtime.

Underwood Falls is very close to Route 9, a mere 50 feet downhill from the road. And yet, despite the waterfall's proximity, it cannot be seen from the road.

A photograph of the waterfall can be seen in Derek Doeffinger & Keith Boas's *Waterfalls of the Adirondacks and Catskills* (2000).

To get there: There are three ways to access Underwood Falls.

West Bank, Approach #1—Head downstream from Split Rock Falls following a faint path that parallels the river. It's not difficult to find, and once you're on it there is no way to wander off in another direction because you are sandwiched in between the river to your right and the highway above you to your left. In less than 0.2 mile you will reach the top of the lower waterfall, where there are limited views.

West Bank, Approach #2—From the parking area, walk down Route 9 for several hundred feet, being constantly mindful of traffic. When you come to a stone guardrail, begin looking for a steep path that leads down to the river. Follow the path halfway down the slope—this can be done fairly easily—for excellent views of the cascade.

East Bank Approach—From the parking area, walk across the Route 9 bridge and follow a wide path down to the base of Split Rock Falls. This is probably as far as most people go. You are not most people, however. Continue following a faint path heading downriver that gradually begins climbing up an embankment above the river. In less than 0.2 mile you will come to a high overlook where Underwood Falls can be seen below. Because you are looking down from a considerable height, expect the waterfall to appear flattened. To compensate, take along a pair of binoculars.

FALLS NORTH OF SPLIT ROCK FALLS

3. NORTH FORK GORGE CASCADES

Having just visited Split Rock Falls, which by all accounts is the most famous waterfall on the Boquet River, we continue upriver to reach the Boquet River's North Fork, starting where Route 73 crosses over the river.

Farther downstream, the merging waters from the North Fork and South Fork were used to power mills in the hamlet of Euba Mills. The hamlet, consisting of several homes, a blacksmith shop, schoolhouse, and sawmill, burned to the ground in 1903. Today, little trace remains.

Several tiny cascades have formed in this rocky, roadside gorge, mostly created by water running over boulders.

To get there: From Underwood (junction of Routes 73 & 9), head northwest on Route 73 for ~1.4 miles. When you come to the bridge spanning the North Fork, park to your right in a long pull-off just after crossing over the bridge (44°06.843'N 73°42.581'W).

Fern Gully. Photograph by John Haywood.

This is a very scenic area, so take a moment to stand on the bridge and gaze upriver. You will find yourself peering into the maw of an impressive gorge that contains tiny cascades, pools, and rapids—all buttressed by towering rock walls. It is a view worth lingering over.

To begin exploring the area, follow a path upstream along the north bank that leads quickly to several overlooks. If you wish you can scramble down to the level of the river for closer, more intimate views. About 300–400 feet upstream from the bridge is a pretty glen called Fern Gully that contains a small cascade.

Views can also be obtained from the south side of the gorge, looking down from a rocky bluff. Unlike on the north bank, however, it is not possible to easily scamper down to the riverbed from these ledges.

As beautiful as this section of the North Fork is, the actual waterfalls begin farther upstream. It's time to grab your hiking stick and start trekking upriver.

4. CASCADE ON TWIN POND OUTLET STREAM

This pretty, 6-foot-high, elongated cascade represents exactly what makes hiking and exploring so rewarding—the presence of an entirely unexpected waterfall.

The lower cascade on Twin Pond's outlet stream.

To get there: From the north side of the gorge several hundred feet from Route 73, follow the path as it leads steeply uphill. This is a vigorous climb that will take several minutes. Once you are at the top, the worst is over. In front of you is a newly cut trail. Previously, hikers had to follow a path along the south side of the North Fork that over the years had become washed out in places—a path that also involved a difficult rock hop across the North Fork to reach the upper falls.

The path that you are following southwest along the north rim of the gorge is roughly 0.6 mile long. It takes you up and down in places, but never descends to the level of the river; hence, any views you have of the river, far below, are fairly distant ones. After 0.6 mile the trail descends to a lower level and quickly comes to a "T." If you turn left, you will immediately come to the North Fork where the original route along the south river bank crossed over from the opposite side. Rather than veering left, turn right (northwest) at the "T," following a well-worn, earlier established trail. In 0.1 mile you will cross over the outlet stream emanating from Twin Pond (which is Round Pond's smaller sister) near a large glacial boulder. Follow the creek downstream for 50 feet and you will come to a lovely, elongated cascade.

5. SHOEBOX FALLS

Shoebox Falls, aka Boquet River Falls, is contained in a massive area of exposed bedrock. Under normal flow the waterfall consists of an 8–10-foot drop into a narrow chasm carved out near the middle of the streambed. You will observe that, just below the chasm, the carved channel, now at a lower level, continues underwater, creating a perfect swimming hole in the summer. In fact, a photo of swimmers at the fall, including a jumper in midair, can be seen in Ben Stechschulte's article "Holy Waters" in the August 2005 issue of *Adirondack Life*.

During times of greater water flow, the entire section of bedrock spanning the stream turns into one enormous, 8–10-foot-high waterfall. Look for scattered branches and a large assortment of debris downstream from the fall just before the point where the river bears right. Some of the debris is entangled in trees. This will give you a sense of just how high the river can get at times.

There are small, slide-like cascades directly above the main fall as well as downstream from its base. Huge boulders are strewn all about. On one visit many years ago we observed a number of artistic looking rock piles that had been erected by hikers on the bedrock—sort of like decorative cairns. The rock formations have long since disappeared, either swept away by the force of a spring freshet or knocked down by "forever wild"–minded hikers.

Shoebox Falls was named for the shape of its swimming hole.

In *Day Trips with a Splash: Northeastern Swimming Holes* (2002), Poncho Doll writes, "Viewed from the top, it's a pile of rock, 15 feet with a chimney through which water drops ten feet into a spectacular hole. The sides [of the hole] are straight and square as a shoe box." It was in Doll's book of swimming holes that I first saw the waterfall referred to as Shoebox Falls. The rectangular-shaped area at the base of the falls is called "Boxcar Swimming Hole."

In *Witness the Forever Wild: A Guide to Favorite Hikes around the Adirondack High Peaks*, Cliff Reiter states that the waterfall drops "into a large, shoe box shaped pool. There are waterfalls and polished rock both above and below the main waterfalls. Rocks are stuck in the crevice in unlikely places." Reiter also includes photos of the waterfall, both wet and dry.

To get there: From the cascade on Twin Pond's outlet stream, continue following the trail west. In less than 0.1 mile you will arrive at Shoebox Falls, on your left (44°06.604′N 73°43.135′W). Take one of three steep paths that lead down to the bank of the river and the falls.

6. UPPER BOQUET FALLS

Twenty-foot-high Upper Boquet Falls is equally as pretty as Shoebox Falls. Like Shoebox Falls, it is also block-shaped with a crevice cut into it by the river, but it is more vertical. On our last visit to the fall, one of our companions scampered right up the face of the waterfall like a mountain goat to reach the top.

Upper Falls on the North Fork of the Boquet River.

To get there: From Shoebox Falls, continue following the main trail south, paralleling the river off to your left. At 0.2 mile you will see a small pond on the Boquet to your left with a tiny cascade at its outlet, its size augmented by beaver activity. There are beautiful views of Dix Mountain to the southwest, particularly if you scamper down to the streambed where the river meanders peacefully across a sandy meadow.

Back on the main trail, proceed south for another 0.1 mile until you come to the point where the trail turns left, descends to the level of the river, and then crosses over it (see "Falls on South Fork of the Boquet"). Instead of turning left here and trekking down to the river, continue straight ahead on a lesser, but still obvious, trail that follows along the ridge line. After another 0.2 mile you will come to Upper Boquet Falls on your left (44°06.349'N 73°43.482'W). It partially faces the trail and is impossible to miss. There is no path leading down the embankment to the base of the fall, so you will have to bushwhack if you wish to see the waterfall up close. My advice is to head down to the base of the fall before the trail passes by it; otherwise you will end up having to scamper through a debris field of rocks, tree limbs, and outwash to reach the fall.

7. BOQUET CANYON & FALLS

Still more falls await over another 1.2 miles upriver, contained in a deep gorge with walls that rise up to as high as 100 feet above the streambed. The main waterfall is called Staircase Falls (named by Loren G. Dobert) and consists of a series of four small cascades stacked staircase-like. Above Staircase Falls is a pretty, split waterfall, dropping into a pool of water. A number of smaller, boulder-choked falls can be seen as well. Near the north end of the gorge, the walls become narrower and the pools formed by cascades increase in number.

To get there: From Upper Boquet Falls, continue northwest. The trail quickly turns into a herd path and eventually disappears completely. Although the hike technically becomes a bushwhack from this point on, you will still be following a trail—only in this case, it's the Boquet River. Just make certain that you keep the river in sight to your left at all times as you head northwest, and then to your right when you return heading downstream, and you should have no problem in not getting lost or disoriented.

The terrain ahead gradually becomes more demanding as you leave the flatlands and begin climbing uphill. After about 1.3 miles and an ascent of 350 feet, you will reach the gorge and its falls, which begin at a GPS reading of 44°06.758'N 73°44.505'W, and end 0.4 mile later at 44°07.085'N 73°44.794'W.

Take note that, once you enter the chasm, you will need to constantly scramble up over and around boulders, talus, and blowdown in order to keep

Staircase Falls is one of several cascades in the Boquet Canyon. Photograph by Loren G. Dobert.

advancing upstream. Some wading may be required as well as rock climbing over waterfalls. There is simply no way to climb out of the chasm once you get started other than to retreat back to where you first entered. This is a hike that is probably best undertaken by a well-prepared party of at least three hikers in the midsummer when less water is flowing and the temperature is warmer. For the rest of us, it may be sufficient to just look down from the top of the gorge.

Interestingly, the canyon is a mere 0.7 mile from Round Pond and Twin Pond (see "Twin Pond Cascade"), but you would really have to know what you're doing to bushwhack that distance over rugged terrain.

8. SOUTHSIDE APPROACH TO RIVER & RAPIDS

There are some who may wish to follow the well-worn trail that goes along the south side of the North Fork from Route 73. This informal path is used by fishermen, hunters, and hikers, and follows close to the river. Along the way you will be accompanied constantly by the chattering sound of the river as it flows over boulders and stones. You will pass a few tiny, 1–2-foot-high cascades.

There are pluses and minuses to taking this path. On the plus side, the path creates a more intimate connection with the river. The downside (one of two) is that recent tropical storms have heavily eroded the path in spots, making it more of a scramble in places.

After following the trail southwest for 0.5 mile, you will come to what I call a decision point. You must choose whether to cross the river here or to continue farther along the south bank. If you choose to cross the river, it will be necessary to rock hop as best as you can, something that is difficult to do in the early spring when the river is running high and the current is swift. Tumbling into cold waters in April or May is a sure way to get hypothermia, leaving you with over a half-mile walk back to your car. One obvious solution is to bring along a pair of water shoes so that you can keep your boots dry, and walking sticks to help with balance. Even so, crossing can be a challenge.

Once you reach the other side of the river, the path is quickly joined by the trail coming in along the north side and you need only follow the directions given earlier to reach the cascade on the Twin Pond Outlet Stream, Shoebox Falls, and the Upper Fall on the North Fork.

If, however, you do not want to rock-hop across the river (and you may very well not want to) then two other choices still remain:

Option # 1—Stay close to the streambed, following along the edge of the North Fork northwest as the river takes a gentle right-angle turn. This will entail stepping on rocks and scrambling up and down the bank at times where the slope meets the river. It is something that, once again, is water-level dependent and not to be attempted during high-water episodes. Also, since the Boquet is constantly eroding the shoreline, what works one year might not work the next. The good news is that you only have to scramble along the riverbed for less than 0.2 mile. After that the path (if that's what you want to call what you've been following) pulls away from the river and takes you to Shoebox Falls in another 0.1 mile.

Option #2—Scramble up a fairly steep embankment just before the river bears right. When you get to the top of the gorge, follow a faint path that heads northwest paralleling the river (which will remain out of sight to your right). In 0.2 mile or so a path will take you back down to the river and Shoebox Falls.

The bad news with both option #1 and option #2 is that if the river is flowing fairly briskly, you may not be able to cross over the North Fork at Shoebox Falls in order to make it to the next waterfall. It is all water-flow dependent.

9. CHASM CASCADE

Chasm Cascade, aka Flume Falls and the Route 73 Flume, is a 6-foot-high cascade that drops into the mouth of a short but impressive-looking chasm. The bedrock at the top of the flume is marvelously sculpted and contoured, riddled with potholes of all sizes and shapes. You will have to be content, however, with looking at the cascade from the top of the chasm. Getting down for a close-up look into the maw of the chasm would be tricky, although some swimmers do so by using a rope ladder. Fortunately, it is possible to view the Chasm Cascade from Route 73 roughly 0.1 mile south of the bridge, the obvious downside here being that the view is a distant one.

Photographs of the waterfall can be seen in Mark Bowie's *Adirondack Waters: Spirit of the Mountain*, Bill Healy's 1986 book *The Adirondacks: A Special World*, and Derek Doeffinger & Keith Boas's *Waterfalls of the Adirondacks and Catskills* (2000), wherein the waterfall is named Flume Falls.

I have one vivid memory about Chasm Falls. Nine or ten years ago Barbara and I were leading a small group of hikers down to see the fall. As I mention below, the embankment leading to the stream is somewhat steep, but not difficult to negotiate if you take your time. Unfortunately, one of our members lost his footing, took a tumble, and came to an abrupt stop, his head inches away from a large rock. He was unhurt, but I shudder every time I think about this incident, for it could have been worse ... a lot worse.

Chasm Cascade can be glimpsed while driving northwest along Route 73.

To get there: You may have already glimpsed this cascade while driving from Underwood to the Route 73 bridge. There is no way to see the waterfall, however, once you reach the bridge, for it lies downstream around the bend following a fault line at the foot of Noble Mountain (2,927').

From the southeast end of the Route 73 bridge (44°06.782'N 73°42.527'W), follow the highway east for a couple of hundred feet, walking close to the guardrail. Then climb over the guardrail and take one of two paths that lead steeply downhill to the chasm and waterfall. Depending upon the path chosen, you may pass by a large, garage-sized boulder. You will quickly come to the chasm where a 6-foot-high cascade drops into its mouth (44°06.773'N 73°42.477'W).

Note: Tiny cascades on a tributary to the North Fork can be seen from the parking area near the northeast end of the Route 73 bridge. Walk over to the edge of the steep embankment and look down. You will see in the stream below two tiny cascades less than 100 feet apart that have formed on a small tributary to the Boquet's North Fork just prior to the confluence of the two streams. This stream rises from the south shoulder of Rocky Peak Ridge. There is probably not a lot to be gained by scampering down the embankment to glimpse the falls close up. They are minor and insubstantial. Still, they are close at hand and part of the Boquet River System.

10. INTERIM STOP

This brief stop along the North Fork of the Boquet provides an opportunity to see a boulder-strewn section of the river. There are no waterfalls on the river here, but don't despair. Diagonally across the road from the pull-off is a mossy, 15-foot-high cascade that runs seasonally. But that's not all. Look closely and you will see that the entire 100-foot-long rock cut, of which the cascade is a part, is alive with water regardless of the time of year. I was so surprised by the amount of water being collected along the side of the road and channeled under Route 73 into the North Fork that I climbed up around the rock cut and hiked uphill to look for an upper cascade. I didn't find one, but I did discover that the whole hillside is one big sponge, making for a very unpleasant surface to walk on.

To get there: From Underwood (junction of Routes 73 & 9), drive northwest on Route 73 for 0.6 mile and turn right into a pull-off (44°06.394'N 73°41.721'W). The river is virtually at roadside, with tiny paths leading to it.

Look across the road at the wall of rock that was obviously created when Route 73 was blasted out, producing a cascade that didn't exist previously.

11. FALLS ON SOUTH FORK OF THE BOQUET

I know there has to be at least one waterfall on the South Fork of the Boquet River, having seen a photograph of it in John Winkler's 1995 book, *A Bushwhacker's View of the Adirondacks*. In pursuit of this waterfall, I followed a well-worn trail for over 1.5 miles to a camping area at the confluence of the two main branches of the South Fork and found nothing. I even bushwhacked an additional 0.5 mile upstream along the left branch to find, once again, nothing of any consequence—just a small stream that seemed to get smaller as it divided and then divided again. Not being completely bushwhacked-out yet, I headed up the right branch for 0.3 mile and again came up empty-handed.

I mention this to you because it exemplifies a harsh reality that I am constantly faced with—not all of my hikes come to successful conclusions. But all is never lost. What this particular hike does offer is an opportunity to follow along one of the two main forks of the Boquet River for a considerable distance and, while you may see no waterfalls, you will eventually reach an oasis of tall conifers with a pine needle floor located at the confluence of the South Fork's two branches. A campsite here is marked by a yellow disk and comes furnished with a fire pit and even a log for sitting by the fire. It is an idyllic setting.

To get there: From Underwood (junction of Routes 73 & 9), drive northwest on Route 73 for 0.2 mile and turn left into a large pull-off a couple of hundred feet before crossing the South Branch (44°06.228'N 73°41.462'W).

From the east end of the pull-off, follow a path up the embankment and then continue southwest on a very well-worn, easy-to-follow trail for ~1.8 miles. For the first 1.2 miles or so you will be surrounded by a forest of deciduous trees and may start to wonder if you are on the right trail—there is no sign of the river once you have gone 0.1 mile (although you do get the sense that the river is not far away). On the mid-fall day I was hiking, a gusty wind kept blowing through the trees, rattling the leaves and creating a noise that mimicked, of all things, the sound of raging waters. Every moment I kept thinking I was coming up to a waterfall—a most disconcerting experience. The cacophony of noise also made listening attentively for the sound of real waterfalls in the distance a virtual impossibility.

Could I have missed a waterfall during the initial part of the hike where the trail kept a significant distance from the South Fork? I suppose it's possible. If I could have heard the river in the distance, I might have been able to distinguish the sound made by a waterfall from the general background sound of the river.

But I digress enough. Let's get back to the main trail. After ~1.2 miles the path leads down to the river and stays by it for the next ~0.6 mile. This trail-side section of the river contains large boulders embedded in a rock-filled streambed but, alas, no cascades—not even a 1-footer!

Eventually, you will come to the confluence of the two branches of the South Fork where a designated campsite can be found.

From the campsite you can continue to follow the trail upstream as it becomes increasingly fainter. It fades in and out, and then essentially disappears after 0.2 mile. From that point on, you're on your own.

There is one other possibility to consider before leaving the general area. As you will see from studying the *Trails of the High Peaks Region* map (if you don't have one, get one, as well as the Adirondack Mountain Club's *Adirondack Trails: High Peaks Region*), a trail that leads from Shoebox Falls to Upper Boquet Falls along the North Fork (see "Upper Boquet Falls") turns sharply left 0.2 mile before Upper Boquet Falls. It then descends to the streambed, crosses the North Branch, and heads southwest, crossing a tributary to the North Branch after 0.2 mile. In another 0.7 mile the trail begins to parallel the South Fork of the Boquet, on your left, roughly 3.0 miles upriver from Route 73. It's possible that this upper section of the South Fork may harbor a waterfall or two. Now is your chance to get out and have some fun exploring.

FALLS SOUTH OF SPLIT ROCK FALLS

Forging ahead, we now leave the North Fork and South Fork of the Boquet River behind, returning to Route 9 to follow the Boquet River, now fully developed, north toward Elizabethtown. Several of its tributaries will be explored as well.

12. CASCADES ALONG EAST TRAIL TO GIANT MOUNTAIN

There are two significant cascades formed on a small seasonal tributary to the Boquet River that runs past the lower section of the East Trail to Giant Mountain.

The first (lower) cascade is an 8-footer located right next to the trail. It is a popular rest stop along the hike.

The second (upper), a 20-foot-high cascade, consists of several drops. The waterfall is fairly inclined, but not to the point of being a waterslide. Directly above is another 8 feet of inclined bedrock.

Farther upstream where the yellow-blazed trail again veers in close to the stream is a 0.05-mile-long section of 1–2-foot-high drops, very pretty, like a Japanese water garden.

To get there: From Underwood (junction of Routes 73 & 9), drive north on Route 9 for 4.7 miles and turn left into the parking area for the "Giant Mountain Wilderness. East Trail to Giant Mountain" (44°08.985′N 73°37.589′W). The trail was cut in 1968.

From the parking area, follow the yellow-blazed trail as it takes you gradually uphill. At 0.7 mile the trail passes directly by a cascade on your left (44°08.789′N 73°38.219′W). Several hundred feet farther upstream is a second, larger cascade (44°08.817′N 73°38.280′W) that can be reached by either bushwhacking upstream or by continuing up the trail for a couple of hundred feet and then bushwhacking over to it, following an obvious ridge line.

Return to the trail. In less than another 0.1 mile the trail again approaches the stream, where a series of 1–2-foot-high cascades can be seen.

13. FALL IN NEW RUSSIA

This small, 2-tiered, ~10-foot-high cascade on the Boquet River (44°09.736′N 73°36.459′W) has been inaccessible for some time, with private homes on both sides of the falls. Fortunately, all is not lost. The waterfall can be glimpsed from roadside, albeit from a distance. Taking along a pair of binoculars would definitely improve the view.

Although this section of the Boquet River is relatively flat, its waters have seen their share of industrial activity. In 1802 a settler named Rich erected a forge on the river, most likely at the falls. Later, H.A. Putnam established a forge. Also evident were a flouring mill and sawmill. Iron smelted in the forges was excavated from iron pits in the local area. It is said that at one time Essex and Clinton counties supplied nearly one-quarter of all the iron ore mined in the United States.

The hamlet of New Russia acquired its name in 1845 when Col. Edmund F. Williams, the Essex County clerk at that time, came up with the idea of putting two disparate words together after he realized that the town's *new* mining technology was being imported from *Russia*.

To get there: From Underwood (junction of Routes 73 & 9), drive northeast on Route 9 for ~6.0 miles.

From Elizabethtown (junction of Routes 9 & 9N North), drive southwest on Route 9 for 3.8 miles.

From either direction, turn east onto Simonds Hill Road and proceed northeast for ~0.1 mile. If you are visiting in the early spring before there are leaves on the trees, you should be able to see the waterfall in the distance, off to the east. Because of the way the river curves, however, you cannot see it from the bridge just ahead, even though the waterfall is close by.

The land between the road and waterfall is privately owned, so please remain in your vehicle and make no attempt to get closer without permission from the homeowner.

14. RICE'S FALLS & US GORGE

We will detour momentarily to visit a series of cascades on the Little Boquet, aka The Branch (as it is often listed on maps)—a major tributary to the Boquet River. The cascades and gorge are collectively known as "US," the exact meaning of which has become obscured over time. Some conjecture that a family named Us was once involved with the falls. Others contend that US was a nickname given to the gorge by local youths, perhaps referring to their collective activities as "us." The fact that there is 2,855-foot-high US Mountain nearby in the Jay Mountain Wilderness Area leads one to wonder if the former might be the more correct conjecture.

The Little Boquet is a medium-sized stream that rises on the east shoulder of Knob Lock Mtn. (~3,189') and flows into the Boquet River just east

Rice's Falls is named after Amos Rice, a nineteenth-century mill operator.

of the center of Elizabethtown. Significant waters from Slide Brook, Falls Brook, and Jackson Brook feed into it, adding their power to create a stream that flows year-round. Rushing through Elizabethtown, the Little Boquet immediately disappears into the Boquet River, with the 0.05-mile-long US Gorge being the river's last hurrah as it makes its way down to the Boquet.

At the head of the gorge is a breached dam. From the dam, the stream produces several small cascades and then plunges over the main, dammed fall (Rice's Falls) where a fairly intact sluiceway on the south bank can not only be seen, but walked through. This is one of only a very few instances where a reasonably intact, enterable sluiceway has survived.

Another example that comes readily to mind is Bartlett Falls in Bristol, Vermont (which I wrote about in *Vermont Waterfalls: A Guide*). In most instances abandoned sluiceways and penstocks end up being physically erased over time or reduced to unrecognizable shapes by the erosive power of the stream. A photo of the mill by Rice's Falls can be seen in a 1998 booklet entitled *Elizabethtown, NY: Bicentennial Celebration, 1798–1998*.

A fairly intact sluiceway can be seen at the top of the falls.

There is a long history of industrialization on the Little Boquet. Around 1792 John B. Roscoe and Stephen Roscoe built the first sawmill on the river by the falls. When Amos Rice erected his sawmill in 1814, the falls then became known as Rice's Falls. This name still lingers on vaguely, preserved by old postcards that show a robust, dammed waterfall labeled "Rice's Falls."

In the 1830s, two forges were constructed on the river between Rice's Falls and the Twin Bridges site on Upper Water Street. The forge closest to Rice's Falls was operated by Deacon Levi Brown; the one just downstream was called the Eddy Forge. Mills were also erected by Jonathan Steele and C.N. Williams & John & Lewis Lobdell.

In 1898 Charles M. Wood used the falls to generate hydroelectric power for the village. The powerhouse was located just below the site of the old Twin Bridges. At some point the hydroelectric plant ceased operations, perhaps a victim of the great flood of 1924. An attempt was made again in the 1980s to use the falls for power generation, but the project was abandoned before completion, leaving behind a cement dam at the top of the gorge that today is breached and slowly being eroded away.

Elizabethtown, aka E-Town, was named after Elizabeth Phagan Gilliland, the wife of William Gilliland, an early settler, surveyor, journalist, and landowner and perhaps one of the most famous men to have inhabited this region. Waterfall enthusiasts will be interested to know that William Gilliland reputedly was the first European to "discover" Ausable Chasm, coming upon it in 1765 (see "Ausable Chasm"). Ironically, although Gilliland "discovered" the chasm, he never traveled far enough upstream along the rim to have seen Rainbow Falls, the chasm's most awe-inspiring feature.

To get there: From Underwood (junction of Routes 9 & 73), drive northeast on Route 9 for ~10.0 miles to Elizabethtown. Turn left onto Route 9N and head north for 1.0 mile. Just before you reach Lord Road on your right, turn right into a large area paralleling the road (44°13.256'N 73°36.849'W). The main fall and sluiceway are located near the downstream end of the gorge (44°13.293'N 73°36.797'W). The path leading to them starts off opposite the driveway to a house across the street. The land is presently not posted, but this could change in the future.

15. SILVER CASCADE

In 1919 Edward Flammer purchased a piece of property northwest of Denton Pond on a small but pretty stream called Silver Cascade Brook. After building his summer home, he named the camp "Singing Waters" because of the sound produced by the brook. The name probably appealed to Flammer who, being in the motion picture industry during the silent era, appreciated the one thing that movies at that time lacked—sound. The brook was called Silver Cascade Brook because of a small, 5–6-foot-high cascade contained in a glen. Seneca Ray Stoddard took a photograph of this cascade that can be viewed in the Sunday, June 19, 2011, issue of the *Adirondack Almanac* available online at adirondackalmanack.com. A postcard photograph of the waterfall, entitled "Singing Waters," can be seen in Scherelene L. Schatz's 2008 book *The Adirondacks: Postcard History Series*. Years later, Flammer's home was converted into a restaurant, only to revert back to a residence in 1956. At some later point the home was destroyed by fire.

If you attempt to find Silver Cascade Brook by name today, you are not likely to succeed. Silver Cascade Brook is now called Barton Brook, and "Denton Pond" no longer shows up on maps. Fortunately, a street named Silver Cascade Way (off Route 10 in Elizabethtown) does appear on road maps, which enables one to hone in on the approximate area where Silver Cascade would be. Unfortunately, however, the land overlooking the cascade remains privately owned.

Silver Falls, aka Singing Waters, was immortalized in an early-twentieth-century postcard.

My wife and I are familiar with the section of Barton Brook that flows past the Stoneleigh Bed & Breakfast, a castle-like home built in 1885 that was converted to a bed & breakfast. We have stayed there a number of times. However, Silver Cascade Way is farther upstream, on the other side of Cross Street, so there is no way to even follow the creek upstream without being intercepted by a road and private property.

16. FALLS NEAR KNOB LOCK MOUNTAIN

No

Several small cascades have formed in a long, deeply cut gorge created by a tiny unnamed tributary to the Little Boquet. These are seasonal cascades; when I visited in early September, the streambed was bone dry. What's amazing is how such a small stream, so insubstantial for most of every year, has been able to carve out such an imposing gorge.

I first read about the cascades in Barbara McMartin's 1987 book *Discover the Northeastern Adirondacks*, where she writes, "The water cascades over fractured rocks for 100 to 150 feet in a series of pools and sheer drops."

To get there: From Elizabethtown (junction of Routes 9 & 9N North), drive west on Route 9N for ~6.6 miles. Park to your left opposite the trailhead to Hurricane Mountain.

From south of Keene (junction of Routes 73 & 9N South), drive east on Route 9N for 3.5 miles and park to your right opposite the trailhead to Hurricane Mountain.

From the parking area (44°12.679'N 73°43.377'W), walk east along the side of the road for 0.1 mile. As soon as you come to the beginning of the guardrail, continue east for another 100 feet and then turn right, heading straight into the woods. If all goes well you should come out to where the waterfall-bearing stream does an elbow turn, changing direction from north to east, parallel to Route 9N. Follow the creek upstream for several hundred feet until you come to the mouth of a gorge. From this point follow along the west rim of the gorge until you begin to see cascades (44°12.545'N 73°43.381'W), most of them created by drops over boulders. In essence, what you will end up seeing is a pretty, cascading stream.

17. WADHAMS FALLS

From Elizabethtown we will continue following the Boquet River until we come to the tiny hamlet of Wadhams where Wadhams Falls—a waterfall of exceptional beauty—can be found. The waterfall is 20 feet high with an 8-foot-high dam at its top. At the base of the fall is a large pool of water, backed up before the channel narrows and drops another 4–5 feet as it races under the bridge. Bedrock, fractured and layered, is virtually everywhere you look.

Wadhams Falls is a 20-foot-high waterfall named after General Luman Wadhams.

Down near the bridge, cement walls and a partially demolished dam are visible, survivors from past episodes of industrialization. Not all has vanished, however. Look downstream from the bridge to your left to see the Wadhams micro-hydroelectric station. Although weathered and battered, it still has plenty of life left in it. The hydroelectric station was constructed in 1904 to furnish power to the iron mines in Mineville and to the Sherman Iron Ore Operation in Withersbee. Later, the Wadhams station provided electricity to Westport, about three miles away. In 1976 the now ailing power station was purchased by Matt Foley, a local businessman. It proved to be a handyman's special. Foley installed a new penstock, rebuilt the dam, repaired the generators, and reopened the station in 1980 to provide power for glass blowing, with excess power sold to Niagara Mohawk.

On the northwest side of the bridge is the Wadhams Free Library, built in 1963, from where excellent views of the fall can be obtained. Some red-colored Adirondack chairs have been strategically placed out in back overlooking the waterfall.

Just downstream from the Wadhams bridge is a mini-hydroelectric plant.

I'm told that Atlantic salmon can be seen swimming around the base of the fall at times, having made it upriver from Lake Champlain before being blocked by the waterfall.

This is Wadhams today. What was it like in the past?

Originally the hamlet was called Coat's Mill, after John Coates. Another name that was used was The Falls.

The first mill at the fall (a gristmill) was erected in 1802 by Jessie Braman and Aaron Felt. In 1819 they constructed a forge, and for a time the hamlet became known as Braman's Mills.

In 1823, Gen. Luman Wadhams purchased a tract of land bordering the fall where Braman and Felt had earlier erected the gristmill and forge. Since both of Braman and Felt's mills had fallen into ruin, Wadhams built his own mills, erecting a sawmill and gristmill. From this point on the hamlet became known as Wadhams Mills, or simply Wadhams. Photos of the sawmill next to the fall behind the present-day library, and the gristmill directly across the street from the library, can be seen in the 1997 book, *A History of Wadhams*.

The name Wadhams has a surprising twist to it. In Old English, Wadhams means "home by the ford." While not identical semantically to "home by the fall," the name does suggest the presence of water, which Wadhams has in abundance.

But there is still more history ...

In 1819 Barnabus Myrick built a forge by the fall, then another in 1825. In 1826 Myrick joined forces with Wadhams to construct a gristmill on the south side of the bridge opposite from where the library is today. In 1865 the property was acquired by Daniel F. Payne, who enlarged the mills and established a sawmill on the west side of the river. Not to be content with those accomplishments, between 1873 and 1875 Payne built two fired forges, and then added two more by 1885. The second channel—probably a sluiceway—that is visible in the river today is where a preponderance of this activity took place.

A map of Wadhams showing where homes and industries were located relative to the waterfall can be seen in a 1993 book, *In the Beginning ... Wadhams. 1820–1993*, compiled by Ethel L. Kozma.

Despite the flurry of activity that characterized the 1800s, by the early part of the twentieth century most of the mills had ceased operations, the end hastened by changing economic times and the structures being repeatedly battered by floodwaters. Sadly, this story of decline seems to be a universal one for most of the mills in the Adirondacks. Wadhams's last mill ceased operations in 1947.

Fast-forward to the twenty-first century. In 2013, when Barbara and I and a group of hikers were at Wadhams Falls on one of our Waterfall Weekend jaunts, a reporter took notice of us and ran a brief article in the *Adirondack Enterprise* about what we were doing. At last! Fifteen years of leading waterfall hikes in the Adirondacks, and finally some recognition.

A photograph of Wadhams Falls can be seen in Derek Doeffinger & Keith Boas's *Waterfalls of the Adirondacks and Catskills* (2000).

To get there: From Elizabethtown (junction of Routes 9 & 9N South), drive east on Route 9N for ~4.3 miles. As soon as you drive under the Adirondack Northway, turn left onto Route 59 (Youngs Road) and proceed northeast for 2.5 miles. When you come to Route 8, turn right and then immediately left onto Route 22 where a bridge crosses the river. Wadhams Falls is just upstream (north) from the bridge (44°13.800'N 73°27.669'W).

From the Adirondack Northway (I-87), take Exit 31 for Westport and Elizabethtown and head east on Route 9N for 0.05 mile. Then turn left onto Route 59 (Youngs Road) and follow the directions given above.

Note: A small cascade on the Boquet River (44°14.894′N 73°30.058′W) approximately 6.0 miles upriver from Wadhams Falls, is located virtually next to the Adirondack Northway, just downstream from the Northway bridge. In *Adirondack Canoe Waters: North Flow* (1987), Paul Jamieson & Donald Morris write that the cascade is a "Class II-III drop of 6 feet in 30 yards." The waterfall is accessible to paddlers, but otherwise appears to be landlocked by a private residence.

18. MERRIAM FORGE FALLS

Merriam Forge Falls, aka Little Falls (little, that is, when you compare it to the size of Wadhams Falls), is a scenic cascade (44°14.651′N 73°25.676′W) located in an area that was once heavily industrialized. You wouldn't know it today, though, for at first glance the area seems remote and unvisited. The dead give-away is the railroad track that passes by the waterfall only dozens of yards away.

The falls are 5–6 feet high and fairly broad. The bedrock forming the waterfall is cut diagonally by a 20-foot-long flume through which most of the river passes. During the spring the entire width of the bedrock is involved, turning into one long and wide waterfall. From the falls, the river then passes through a 100-foot-long gorge with 10–15-foot-high walls.

Merriam Forge Falls is named after William P. Merriam and P.D. Merriam, who erected a forge at the falls in 1825. Theirs was a fairly substantial mill containing three fires. Eventually the forge fell into disuse and finally vanished completely when its ruins were obliterated in the great flood of 1897.

Today, this stretch of land along the Boquet River is maintained by the Adirondack Land Trust and is primarily used by anglers.

To get there: From Wadhams (junction of Routes 22 & 10) drive northeast on Route 22 for 1.8 miles. Turn right onto Merriam Forge Road and proceed downhill, heading south, for over 0.2 mile. As soon as you cross over railroad tracks, turn left onto a dirt road and follow it north for several hundred feet to a parking area (44°14.522'N 73°25.794'W). If you cross over the Boquet River, you have gone too far.

From the parking area, follow a continuation of the dirt road north for 0.1 mile, paralleling the river to your right and the railroad tracks to your left. When you reach a point where the trail ends, you used to be able to walk up to the railroad track and then continue north along the track, being mindful of oncoming trains, for a hundred feet, and then walk right down to the falls.

Merriam Forge Falls was once heavily industrialized.

The area is now posted by the Canadian Pacific Railway Police Service, and trespassing is not allowed. This posting is of fairly recent origin. It was only three years ago that we led a hike to the falls without the site being posted. It seems likely that the posting was the result of Canadian Pacific Railway's concern that a hiker might be struck while walking along the tracks. This is not a totally farfetched scenario. Accidents of this nature do happen now and then for the simple reason that hikers sometimes fail to hear the approach of the train from behind.

It reminds me of an incident that occurred to me and Barbara at the east entrance to the Hoosac Tunnel in Massachusetts when we blithely followed the tracks into the tunnel's interior for a good 300–400 feet. All the time we were aware of a lamplight in the distance, but it seemed unvarying and stationary. We just assumed that the light was coming from a handcar being used by railroad maintenance employees. After looking around for a few moments, we walked back out of the tunnel and were heading toward our car, which was parked in an area by the Deerfield River. Suddenly, without warning and to our utter astonishment, a train came barreling out of the tunnel doing 60 MPH and flew right past us. We never heard a sound or felt a vibration until the train was upon us. Fortunately, we were to the side of the tracks. What we hadn't taken into account was that the Hoosac Tunnel is nearly 5 miles long and straight as an arrow. You can probably almost see from one end to the other. What we thought was a stationary light was actually the headlamp of the onrushing train that only seemed motionless because of its distance away. A close call? Well, it was closer than I would have liked, that's for sure.

Getting back to Merriam Forge Falls, the only legal way of proceeding would seem to be to bushwhack either along the south side of the Boquet, a distance of probably 0.2 mile, or to try to follow the shoreline of the river down from the end of the trail without going up by the railroad tracks.

Note: Another small cascade on the Boquet River (44°18.259'N 73°24.285'W) is located in the tiny hamlet of Bouquet, approximately 6.0 miles downriver from Merriam Forge Falls. It is described by Paul Jamieson & Donald Morris in *Adirondack Canoe Waters: North Flow* (1987) as a set of "complex ledges, cascades, and rapids." The cascade's best feature is that it can be seen from the Jersey Street (Route 12) bridge that spans the river just upstream from the fall.

19. WILLSBORO FALLS

Willsboro Falls, aka Milltown Falls in the town's early days, or simply "The Falls," is the last waterfall on the Boquet River, located only a short distance upriver from Lake Champlain. The cascades begin downstream from where a hydroelectric dam previously spanned the river, starting with a 75-foot-long, natural waterslide that drops 5 vertical feet, followed by a 2-foot-high ledge cascade, then another slide dropping 6 vertical feet over a distance of 30 feet, and finally ending with a series of rapids. To be sure, Willsboro Falls are not flashy. They resemble something you might expect to see in a waterslide park.

The first dam was built in 1813 and then rebuilt in 1890 to furnish hydropower to a paper mill, gristmill, and iron forge located downstream. It was demolished in 2015 by engineers, much to the consternation of many townsfolk who had grown accustomed to its presence.

The hamlet of Willsboro was settled in 1765 by William Gilliland, an early landowner whose name and exploits permeate the region. Although the town was initially called Milltown because of its sawmill, the name later changed to Willsboro in honor of its famous settler, William (Will) Gilliland, a pioneer colonial land developer.

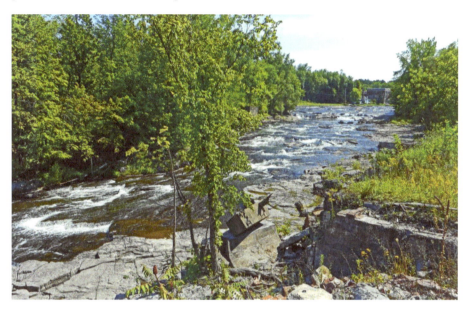

Willsboro Falls is the Boquet River's last hurrah before flowing into Lake Champlain.

In 1801 Levi Highby and George Troop erected a forge at the falls. They crafted mill irons and anchors that weighed from 300 to 1,500 pounds each. Most were delivered to Troy, New York.

In 1884 the Champlain Fibre Company was established on the west side of the river, producing wood pulp for paper-making. Later, the name changed to the Willsboro Pulp Mill. The mill was a major employer for the town until its operations ceased in 1964. Eventually the property was deeded to the town of Willsboro by the Georgia-Pacific Corporation in 1966.

The first bridge to span the Boquet River upstream from the falls was built by Platt Rogers in 1790. There have been others since then.

To get there: From Wadhams (junction of Routes 22 & 10) follow Route 22 northeast for ~14.0 mile to Willsboro. Just before crossing over the bridge spanning the Boquet River, turn right onto Gilliland Lane, heading north. In 0.1 mile you will see a parking area on your left with two park benches on a cement slab that overlooks the river (44°21.999'N 73°23.448'W). From this spot, views looking downstream over the cascades can be obtained.

Continue north down Gilliland Lane for another 0.1 mile and turn left into Gilliland Park (44°22.084'N 73°23.486'W). From there, walk along a road-like path for 75 feet that leads out to an embankment overlooking the river where a kiosk provides information about the site across the river. From either end of this overlook, walk down a flight of log steps to the diked shore of the river to obtain views looking upriver at the cascades. On the day of our visit, there were over a half-dozen anglers fishing along the shoreline. It has been, and continues to remain, a favorite site for fishing, particularly in the spring when salmon are swimming upstream, and in the fall when they spawn.

Should you continue driving down Gilliland Lane from the park for another 0.3 mile, you will come to a boat launch, but there are no more cascades to be seen.

A second option for viewing the falls is to drive over the Main Street Bridge spanning the Boquet River and then turn immediately right onto Mill Lane, heading north. In 0.3 mile you will come to a cul-de-sac with good views upriver of the cascades. The cul-de-sac occupies the site of a former paper mill that operated from 1884 to ~1964.

II: NORTH HUDSON

The earlier Elk Lake Dam.

North Hudson, formed in 1848 as an offshoot of Moriah, got its name from its location near the north end of the Hudson River. In the early years its economy was tree-based—first tanneries, then lumbering. From 1952 to 1998, Frontier Town (a Wild West family-theme park) was the community's driving economic force.

North Hudson conveniently lies at the southeastern edge of the Adirondack High Peaks.

There is currently serious talk underway about New York State creating a multimillion-dollar "Gateway to the Adirondacks" Visitor Center off of Exit 29 of the Adirondack Northway. It would be built on the former site of Frontier Town. If this plan is carried out, one of its goals will be to divert hikers from the High Peaks Region to other less visited areas, such as the Finch, Pruyn & Company lands to the west as well as some of the hiking areas mentioned in this section of the book.

20. BLUE RIDGE FALLS

Blue Ridge Falls (43°57.338'N 73°46.959'W) is formed on The Branch—a medium-sized stream that rises from Elk Lake and flows into the Schroon River southwest of North Hudson. The falls have formed where the stream reaches a huge area of blocks and ledges cut by the river as it drops some 20–25 feet over a distance of a hundred feet. Although no individual drop is noteworthy in itself, the combined collective sight of so many bursts of falling water is both scenic and impressive.

Roadside view of Blue Ridge Falls. Photograph by John Haywood.

Directly across from the overview is the Blue Ridge Falls Camp Site, located at 3493 Blue Ridge Road #84, North Hudson 12855, (518) 532-7863. Their Web site is blueridgefallscampsite.com.

A section of the Blue Ridge Road was once part of the 1841 Carthage Road linking Crown Point with the Black River. Although the Adirondacks were still in wilderness early in the nineteenth century, some rough roads did provide avenues through what would otherwise have been an impassable landscape.

There has been a fair amount of industrial activity on The Branch. In his 1885 book, *History of Essex County*, H.P. Smith writes, "The attempts at working iron in this town comprise the forge built on the Branch about a mile from the hamlet of North Hudson by Jacob Parmerter." This would place the mill over one mile downstream from Blue Ridge Falls. Smith also writes, "Another tannery was built by Sawyer & Mead about three miles west of the hamlet of North Hudson on the west branch of the Schroon," which places this enterprise close to or even at Blue Ridge Falls.

Lest hikers get this falls confused with another that sounds similar, there is a Blue *Ledge* Falls in the Adirondacks that has formed on the upper Hudson River just upstream from a well-visited spot called Blue Ledge. It is one of the river's more notorious rapids/cascades

To get there: From the Adirondack Northway, take Exit 29 for North Hudson and proceed west on Blue Ridge Road for over 2.3 miles. Turn left into a pull-off where an exceptional roadside view of Blue Ridge Falls can be obtained (43°57.395'N 73°46.887'W).

Less than a hundred feet farther west on Blue Ridge Road is a second pull-off, also on your left. From there a wide path leads to a narrower path on your left that takes you down to the falls. The path is not posted at present, but access to the falls may be restricted. The grand view, "the big picture," is from the roadside pull-off.

21. THE FALLS

The Falls, also called West Inlet Falls and Wagon Wheel Falls, is formed in a notch between Blake Peak (3,960') and Nippletop (4,620') on the West Inlet, which flows into Elk Lake. The waterfall is approximately 30 feet high. When there is a lower volume of water, the fall strikes an inclined ledge halfway down and from there proceeds to run down the shelf into a pool at the base of the fall. My best estimate for the waterfall's location is a GPS reading of 44°04.660'N 73°49.926'W at an elevation of around 2,400 feet.

The Falls is accessible from Elk Lake Lodge, a family-run, multigenerational lodge that lies at the center of the 12,000-acre private Elk Lake–Clear Pond Forest Preserve on land protected by a New York State Conservation Easement. It contains over 40 miles of private trails, one section of which leads to the waterfall. The main lodge dates back to 1904. In an issue of *Outside Magazine*, Elk Lake Lodge was rated as one of the ten best wilderness lodges in North America. For more information on Elk Lake Lodge, write to them at 1106 Elk Lake Road, North Hudson, NY 12855, call (518) 532-7616, or consult their Web site, elklakelodge.com.

Elk Lake is the centerpiece of the preserve, encompassing an area of 600 acres. It contains 30 islands, plus a dam at the lake's south side. The lake was known as Mud Pond until 1873 when Verplanck Colvin, in his survey report *The Adirondack Wilderness*, changed it to Elk Lake to honor the great Alpine animal.

The Falls is a striking, 30-foot-high waterfall formed on West Inlet. Photograph by Mike Sheridan.

To get there: From the Adirondack Northway (I-87), take Exit 29 for North Hudson and drive west on Blue Ridge Road for ~4.1 miles. Turn right onto Elk Lake Road and proceed north for ~5.0 miles to Elk Lake Lodge (44°01.356′N 73°49.748′W).

The real adventure now begins for guests staying at the lodge. The trek to The Falls is a combination of water and land. Using a canoe, kayak, or rowboat provided by Elk Lake Lodge, paddle across the lake to Wagon Wheel Landing on the northwest side, a distance of roughly 3.0 miles. Take note that in order to prevent the introduction of invasive species to the lake, you are not allowed to use your own kayak or canoe.

Debarking from Wagon Wheel Landing, the land trek to the waterfall involves a hike of ~2.6 miles. Follow the yellow-blazed Wagon Wheel Trail northwest, crossing over Virginia Creek via a bridge at 0.5 mile, then a second bridge over an unnamed stream at 0.7 mile. Continue north, passing by the orange-blazed Beech Ridge Trail to your right at 0.9 mile. At 1.8 miles you will come to the Upper Camp clearing. At 2.4 miles West Inlet Brook is reached. Cross shallows and follow the brook upstream for 0.2 mile to The Falls.

Private trail maps are available to guests of Elk Lake Lodge.

22. WEST MILL BROOK FALLS

West Mill Brook Falls are formed on West Mill Brook—a medium-sized stream that rises from the southeast shoulder of Macomb Mtn. (4,405') and flows into a tributary of the Schroon River. Macomb Mtn. was named for General Alexander Macomb, whose victory over the British at Plattsburg in 1814 was a critical battle for America.

The hike follows an old road/trail along West Mill Brook that was once part of the Cedar Point Road that linked the Schroon River valley and Clear Pond, south of Elk Lake. One can only imagine what that road must have been like as it headed up to and around these huge mountains. During the mid-to-latter part of the nineteenth century, the woods in the West Mill notch were extensively harvested. This accounts for the present condition of the road/trail, which is still well-defined.

There are two separate sections of waterfalls on the stream, and they are separated by less than 0.1 mile. Both are worth exploring.

Site One contains a 200-foot length of exposed bedrock with multiple drops and small cascades. Large boulders abound. Several of the cascades are at least 3 feet high, which may not sound like much, but put them into a scenic area like this and the result is pure magic.

Site Two contains a phenomenon rarely observed in the Adirondacks—a plunge fall. At the top of the waterfall, the stream is funneled through a large pothole and then plunges 10 feet into a pool of water below. During times of high water, the section of bedrock spanning Mill Brook just downstream from the waterfall may back up the water sufficiently to produce its own cascade and cause the plunge pool to rise up to nearly the height of the plunge fall. As is true for site #1, there is much exposed fractured bedrock.

If you are visiting later in the summer, take note of the west bank, where a dry streambed arcs around the plunge fall and then back to the stream. A 6-foot-high block of rock near the end of this streambed becomes animated during times of high water flow, producing an additional 6-foot-high waterfall.

To get there: Driving north on the Adirondack Northway (I-87), take Exit 29 for North Hudson and turn right onto Blue Ridge Road, heading east and going past the abandoned ruins of Frontier Town, on your right. After 0.3 mile turn left onto Route 9 and proceed north for ~5.5 miles.

Driving south on the Adirondack Northway, take Exit 30 and head south on Route 9 for 4.1 miles.

A number of small-to-medium-sized cascades have formed on lower West Mill Brook.

At the sign for the "Dix Mt. Wilderness Area. West Mill Bk Access to Dix Wilderness Boundary" (44°01.416'N 73°41.275'W), turn west onto a dirt road that leads downhill to a tributary of the Schroon River in 0.1 mile. You can either park off-road here and proceed the remainder of the way on foot (which is what I did to lengthen the hike) or continue by driving across the stream, which is very doable, and then heading west on a well-maintained dirt road. Take note of repeated orange-colored signs along the road that admonish, "Public Right of Way across private land. Do not leave this road."

There is a fair amount of activity going on back here, and you will see industrial machinery scattered about. After 0.7 mile you will come to the Adirondack Northway's two underpasses, which are wide enough to allow cars to drive through unimpeded. After driving or walking under the Northway, continue south for ~0.3 mile until you come to a barricade marking the point where the drivable portion of the road ends and state land begins (44°01.509'N 73°42.286'W). West Mill Brook will be visible to your right at this point. If you have driven in, park your car in the large area before the barrier.

Continuing on foot, you will come to a side path on your right in 0.05 mile that leads down to the stream and then across it via a swinging footbridge to Erick's Camp. While an interesting side diversion, there are no cascades to be found here, so continue straight ahead on the main trail. In less than 0.2 mile, just before you reach a rusted, yellow barrier, a shallow gorge is visible to your right (44°01.632'N 73°42.458'W). Bushwhack down to the gorge, which is less than 100 feet away. It is not difficult to do. A number of small cascades can be seen that are contained in an impressive section of exposed bedrock that extends for approximately 200 feet. It is a very scenic spot.

Return to the main trail. Walk past the yellow barrier and proceed west for another ~0.1 mile. Look for a faint path to your right that starts down toward a huge area of exposed bedrock less than 100 feet away (44°01.708'N 73°42.550'W). This is where the main waterfall is located.

Return to the main trail again, which continues northwest for several miles farther, fading out partway up the mountain. Other falls are listed on topographical maps beyond where the trail ends, involving what I would assume to be a fairly arduous bushwhack or semi-bushwhack of 2.0–3.0 miles. This is one trek that I have not taken. An account of the hike, however, was written by Barbara McMartin in an article "The Other Niagara Falls" in the *Adirondack Life 1999 Annual Guide to the Adirondacks*. Two cascades south of Macomb Mtn. are described, including a photograph of one of the falls that looks as large as Beaver Meadow Falls.

A photo of the upper West Mill Brook Falls on the east side of Macomb Mtn. can be seen in John Winkler's 1995 book *A Bushwhacker's View of the Adirondacks*.

Topographical computer software shows the waterfall on the shoulder of Macomb Mtn. to be at a GPS coordinate of around 44°02.039'N 73°45.518'W.

The other waterfall—McMartin's Niagara Falls, playfully so-called because it is formed on Niagara Brook—lies farther south.

23. LINDSAY BROOK FALLS

A number of waterfalls of various sizes and shapes have formed on Lindsay Brook, a medium-sized stream that rises east of Spotted Mtn. (3,445') and flows into a tributary of the Schroon River. In the March/April 2001 issue of *Adirondac*, Jean Halcomb writes about hiking along Lindsay Brook where the "Next point of interest was a huge thundering waterfall, one of so many over the course of the day that after a while we didn't even bother to stop and look." This description is irresistible to a waterfall seeker.

For Jean Halcomb and her friends, the hike proved to be a fairly demanding bushwhack when they did it in 2001. Today it is much harder. A sign at the Northway underpass states, "Notice—No trail maintained beyond this point. Do you have a compass, maps, matches? NYS Conservation Department." If that's not enough to discourage the faint-hearted, New York State closed the trail in 2007, making it even less inviting.

Why, then, do I mention the falls on Lindsay Brook? Perhaps I'm being overly optimistic, but with the need to siphon off more hikers from the heavily traveled High Peaks Region, is it unreasonable to expect that maybe this trail will be reopened at some point in the future? Time will tell.

Sharp Bridge, near where the trail begins, is historically significant. A state historic marker at the northeast end of the bridge brings attention to the fact that the campground was one of the first two built on Forest Reserve Land in 1920 by the Conservation for Public Recreation. The other campground was built south of Wells on the Sacandaga River around the same time.

To get there: From the Adirondack Northway (I-87) take Exit 29 for North Hudson and drive east on Blue Ridge Road for 0.3 mile. Turn left onto Route 9 and head north for ~7.0 miles to Sharp Bridge. Just before crossing Sharp Bridge, turn left into a pull-off by the trailhead.

Driving south on the Adirondack Northway, take Exit 30 and proceed south on Route 9 for ~2.6 miles. As soon as you cross over Sharp Bridge, turn right into a pull-off.

A sign at the pull-off states, "Hammond Pond Wild Forest Trail to Lindsay Brook. 0.9." Next to it is a sign that emphasizes, "Trail closed."

From the parking area (44°02.606'N 73°40.714'W) follow a 0.9-mile-long trail (an old abandoned road) north as it takes you to, and under, the Adirondack Northway. You will have to walk through a marshy area near the beginning of the hike and then negotiate blowdown that frequently lies strewn across the trail. Occasionally you will see old red disks on trees from days when the route was an open hiking trail. When I recently hiked it, I was struck by just how quickly an unmaintained trail can turn wild. I was left with a newfound appreciation and respect for all the vital and hard work that trail maintenance crews do for us.

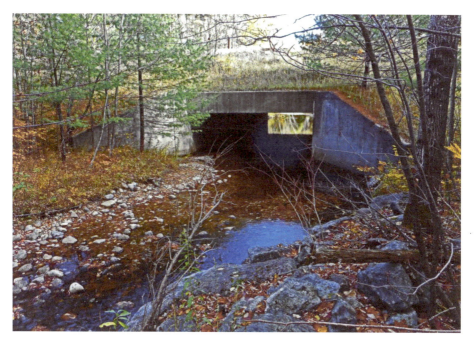

Multiple falls on Lindsay Brook are only a bushwhack away upstream.

Upon reaching the Adirondack Northway, whose distant hum accompanies most of the hike, walk through two consecutive large tunnels under the superhighway and out into a wild forest. From this point on there is no official trail and you should not go any farther unless you are an experienced hiker, joined by several equally competent companions, and able to bushwhack reliably through wilderness.

There is, however, a *faint* informal trail that takes you north between the hilly base of Saunders Mtn. and the Adirondack Northway. (Maybe deer are maintaining it now.) I followed the trail north for 0.2 mile and came out to where Lindsay Brook flows through a culvert under the Adirondack Northway (44°03.558'N 73°41.087'W). I stopped at this point, my goal being simply to see if there was a viable way to get to Lindsay Brook. There is. If I needed to continue upstream along the north side of the brook, I could have easily scampered up and over the Adirondack Northway culvert without actually going up to the highway. It was late in the day, however, and I wasn't about to undertake the Lindsay Brook hike without a couple of companions and more daylight.

One unexpected find was coming across a stubby, triangular-shaped, concrete marker set in the ground with the initials NY on it. I encountered it just before Lindsay Brook. An old highway marker, perhaps?

24. CROWFOOT BROOK CASCADES

Three small cascades have formed on Crowfoot Brook, a small stream that rises from Crowfoot Pond and flows northwest into Deadwater Pond. The falls are located just downstream from a footbridge that crosses Crowfoot Brook. The uppermost cascade is very distinctive looking, dropping 10 feet down through a mini-chasm. On both sides of the stream, the bedrock has fractured and collapsed inward, leaving behind a fairly untouched, oblong centerpiece of bedrock with the stream going around each side of it.

Fifty feet downstream is a second cascade, about 6 feet in height. Check out the orientation of the bedrock, which is tilted nearly 90 degrees.

Just below is the third cascade, a 6–8-foot-high waterslide that extends for over 20 feet.

To get there: From the Adirondack Northway (I-87) take Exit 30 and drive south on Route 9 for ~200 feet. Turn left onto Tracy Road (Route 6) and proceed east for over 1.6 miles, turning right at a sign that states, "Hammond Pond Wild Forest. Crowfoot Pond 2.5 miles." Follow a dirt road in for 100 feet until you come to a sizeable parking area (44°04.390'N 73°37.868'W).

From the parking area, walk across Crowfoot Brook via an impressive footbridge that was erected within the last three years, replacing an earlier one that was washed out. Turn left and begin following the yellow-blazed Crowfoot Brook Trail (an old wagon road) east. You will encounter four more footbridges before reaching Crowfoot Pond; one of the footbridges crosses over a tiny tributary called Newport Brook (no, there are no cascades here). When you come to the first footbridge, at 0.7 mile, look for the cascades to your left just downstream from the footbridge (44°04.089′N 73°37.209′W). All three are close to the trail but will require a slight scramble if you wish to get closer.

Small falls are encountered along the 2.5-mile trek to Crowfoot Pond.

On the day of my visit, there was barely any water flowing. Although it's unlikely that I missed any other cascades along Crowfoot Brook, there is no way to know for sure given the minimal flow of water and the fact that the brook was, literally, not speaking to me. (There's a sound that waterfalls make that is different from the regular background chatter of a stream).

The trail comes to an end at Crowfoot Pond in 2.5 miles, where a small clearing gives you a chance to look out across the pond. My arrival was accompanied by what sounded like the report of a cannon or rifle. In actuality a beaver had slapped the water with its tail, perhaps as a warning to other beavers in the area. I gazed across the pond, but the wily creature had disappeared.

All was not lost, however. One hundred feet before the end of the trail is a tiny tributary entering from the left, where I believe a 10-foot-high, seasonal cascade is produced in the early spring (44°03.538'N 73°35.527'W).

25. ASH CRAFT BROOK FALLS

Thousands of automobile drivers heading south on the Adirondack Northway (I-87) from parts farther north have undoubtedly glimpsed in passing a small cascade to the right (west) side of the highway several hundred yards from the road, roughly 0.5 mile before Exit 30. The waterfall is Ash Craft Brook Falls, a double cascade formed on Ash Craft Brook (44°04.930'N 73°39.177'W), which rises from Ash Craft Pond, a small body of water that lies next to the Adirondack Northway about 2.0 miles north and is a tributary to the Schroon River. The upper cascade looks to be ~6 feet in height. The lower one, perpendicular to the first, looks to be 6–8 feet high and significantly inclined.

You will have to look quickly to see Ash Craft Brook Falls as you drive along the Adirondack Northway.

Who could ask for a better secluded waterfall!

There is one catch. Ash Craft Brook Falls can only be accessed from the Adirondack Northway, a superhighway where cars are not permitted to pull over and stop except at rest areas or in an emergency. The end result, then, is that while you can catch a glimpse of the waterfall, you cannot stop to hike up to it. While Ash Craft Brook Falls teases, it never fully pleases.

To get there: Driving south on the Northway between Exits 31 and 30, look for a green sign on your right that states that Exit 30 is one mile away. Clock 0.5 mile from the sign and watch for the cascade to appear on your right, upstream from a marshland (44°04.892'N 73°39.106'W). Look quickly; the window of opportunity to see the waterfall is fleeting.

Some hikers cannot resist the impulse to impose their will on nature. Downstream from Shoebox Falls.

Chapel Pond—a roadside gem.

There are a number of cascades along Route 73 in the general area of Chapel Pond that lie close to the road or involve a short bushwhack. Few of them have formal names, and the streams upon which they are formed generally remain nameless as well.

Chapel Pond is a gem of a mountain lake located between the towering cliffs on Round Mountain's shoulder and the Washbowl cliffs of Giant Mountain. The pond, including the two Ausable Lakes, is what remains of post-glacial Lake Keene. To keep fishermen satisfied, the Essex County Fish Hatchery regularly stocks the pond with several hundred two-year-old rainbow trout.

Occasionally, paddlers can be seen plying the waters in canoes or kayaks. Inevitably, there are rock climbers during all seasons of the year.

A road between Underwood (near the junction of Routes 73 & 9) and Keene Valley through the Chapel Pond Pass was completed by 1846, possibly even earlier according to one source, but it wasn't until modern times that Route 73 developed into the scenic highway that we know today.

26. HIDDEN CASCADE

Hidden Cascade consists of a two-tiered seasonal waterfall—a 15-foot-high cascade that drops into a shallow pool and, above it, coming in perpendicularly, an 8-foot-high cascade. The small falls are formed on a fairly insubstantial stream that rises from the southwest shoulder of Rocky Peak Ridge (4,420').

Hidden Cascade is one of two hidden waterfalls along Route 73.

To get there: From Underwood (junction of Routes 73 & 9), drive northwest on Route 73 for 2.2 miles (or 1.2 miles from the bridge spanning the North Fork of the Boquet River). Turn right into a pull-off (44°07.358'N 73°43.162'W).

The hike up to Hidden Cascade is short but moderately difficult because it is a bushwhack. From near the center of the parking area, follow a steep, zigzagging path that takes you down to the streambed just downstream from a large meadow containing a proliferation of dead trees. Cross over the stream emanating from the meadow (which admittedly can be a bit tricky depending upon the time of year and water flow) and then head upstream along a tiny tributary that lies very close to the meadow's outlet. After a hundred feet you will come to the stream's lower, 15–20-foot-high cascade (44°07.385'N 73°43.036'W).

To get to the upper cascade, cross over the stream and scamper up its right bank until the 8-foot-high cascade comes into view.

This waterfall is best visited in the early spring when water flow is at its optimum. The only problem, however, is that the stream paralleling Route 73 is also running at full force then and can be difficult to cross.

27. SECLUDED CASCADE

Secluded Cascade is the second cascade that can be found near Route 73 between the bridge spanning the North Fork of the Boquet and the Round Pond parking area. The waterfall is formed on an insubstantial stream that runs over the top of a ridge line, drops 15 feet, and then tumbles through a steeply inclined ravine choked with boulders as though careening through a pinball machine. A small cascade has formed near its base next to a 10-foot-high boulder.

To get there: From Underwood (junction of Routes 73 & 9), drive northwest on Route 73 for 2.6 miles (or 1.6 miles from where Route 73 crosses over the North Fork of the Boquet). Turn right into a pull-off on the side of the road (44°07.621'N 73°43.420'W). During times of high water flow, the lower part of the stream is visible through the woods, but not the upper section where the main cascade has formed.

From the pull-off, follow an obvious path downhill for 40 feet to a stream that runs parallel to the highway. Turn right onto the path and follow the brook downstream for another 40 feet. At this point cross over the stream as best you can and follow a tiny tributary uphill for 80 feet to reach the base of Secluded Cascade (44°07.637'N 73°43.377'W). You will need to climb higher in order to get a picture of the 15-foot drop, which comes in at an angle.

Secluded Cascade is unfamiliar to most hikers.

    This is a much easier cascade to reach than Hidden Cascade, involving only a short bushwhack. On the other hand, the creek forming the cascade is *very* seasonal. Should you visit in July or August, you may find only a dry creek bed for your efforts.

28. TWIN POND CASCADE

The cascade on Twin Pond's outlet stream is a fairly insubstantial one—more like a miniature waterslide. We could make a very convincing argument that few people know of its existence or have any reason to visit it. What's more, the cascade is formed on the outlet stream of a minor body of water a mere 0.05-mile-wide. For these reasons Twin Pond and its tiny cascade are generally ignored as hikers hurry up to Round Mountain (3,100') or the Dix Mountain Range (4,857'), making their way past the northwest side of 23-acre Round Pond—a bucolic body of water that lies 0.1-mile north of Twin Pond and is one of the best brook trout waters in the county according to the New York State Department of Environmental Resources.

The view looking across Twin Pond.

To be sure, Twin Pond is not really a twin of Round Pond no matter how you stretch your imagination. Round Pond is fairly round; hence, its name.

Twin Pond, by contrast, is about one-third of Round Pond's size and shaped more like a bullet.

This tiny cascade is not the only one formed on the outlet stream. A larger cascade located 1.0 mile farther downstream was described in an earlier hike. (See "Falls on the North Fork of the Boquet.")

The tiny cascade on Twin Pond's outlet stream can be a bit underwhelming— but the hike to it isn't.

To get there: From Underwood (junction of Routes 73 & 9), drive northwest on Route 73 for 3.2 miles (or 1.8 miles from the bridge spanning the North Fork of the Boquet) and pull into the parking area for Round Pond, on the west side of the road (44°07.923'N 73°43.918'W).

Walk southeast back down the road for 200 feet; then follow the blue-blazed "Round Pond, Dix Mtn, and Noonmark Mtn" trail south uphill for over 0.5 mile. Once you reach Round Pond, bear left as the main trail goes right, and follow a path that takes you along the east side of the pond for 0.1 mile to a campsite by the south end of Round Pond where a tiny stream flowing out of Round Pond connects it to Twin Pond.

From there the hike gets dicier. A path leads south from the campsite, following close to Twin Pond, but fades in and out and then disappears. You will have to keep your wits about you as you bushwhack 0.2 mile to the south end of Twin Pond and its outlet stream. As long as you keep the pond in sight you will be in no danger of getting lost or disoriented. Finally, follow the outlet downstream for 0.1 mile to the waterslide cascade (44°07.176'N 73°43.875'W).

Several hundred feet farther downstream is an oval area of exposed bedrock roughly 25 feet by 25 feet. The bulge seems incongruous with a stream that is otherwise narrow.

Don't be disappointed if the cascade encountered here is less than what you expected. I'm not promising much. Half of the fun is just getting out and exploring, and I can say with a fair degree of confidence that you will have the place all to yourself. Intimate views of these two mountain ponds will most surely make up for any disappointments incurred.

29. TWIN FALLS

This 30-foot-high seasonal cascade is formed on an unnamed stream that rises from the col between Giant Mountain (4,627') and Rocky Peak Ridge (4,420'). The waterfall got its name from John Haywood, an Adirondack photographer. It consists of several drops on a towering, cascading stream. The two main drops, somewhat similar in appearance, are what give the fall its name.

Downstream from the base of Twin Falls, the creek turns left and parallels Route 73 as it makes its way south down to the North Fork of the Boquet River.

Like Hidden Cascade and Secluded Cascade, Twin Falls is best seen in the early spring. By late summer it has all but disappeared.

To get there: From the parking area for Round Pond (see "Twin Pond Cascade" for directions), cross over to the east side of Route 73 and walk northwest along the side of the road for several hundred feet. Look for a large

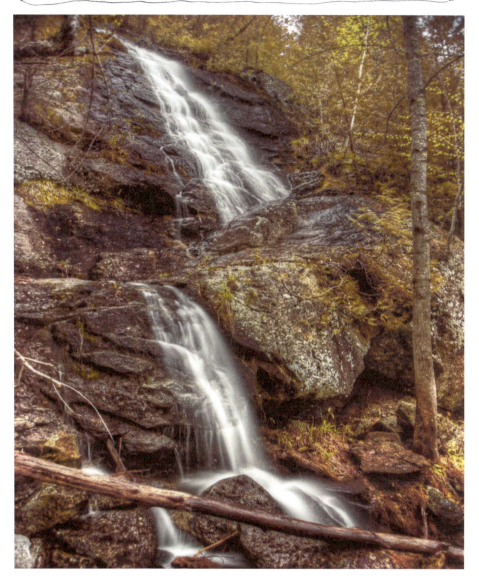

Twin Falls was named for its two, similar-looking cascades. Photograph by John Haywood.

roadside boulder on the northeast side of the road. Then turn right into the woods, following a stream uphill (north) for a hundred feet to reach the base of the cascade (44°08.002'N 73°43.939'W). If you're visiting in the early spring, the cascade can be heard from the road, making it that much easier to locate. The most interesting views of this cascade are from about halfway up.

30. ROCK GARDEN FALLS

Rock Garden Falls seems like an appropriate name for a 15–20-foot-high waterfall that cascades down a nearly vertical, dark-colored rock wall. It is just the kind of cascade that you would expect to see featured in a Japanese rock garden.

Because my wife and I had never noticed this waterfall until fairly recently, my initial thought was that the cascade must be of recent origin, perhaps caused by a realignment of the creek. However, when I scrambled up to the top of the waterfall, all I saw was a very well-defined streambed that looked like it had been there for centuries. So much for that theory!

Rock Garden Falls is like something out of a Japanese rock garden.

Rock Garden Falls is formed on a fairly insubstantial unnamed stream that rises from the lower, south shoulder of Giant Mountain (4,627') and flows into Chapel Pond a short distance to the west.

To get there: From Underwood (junction of Routes 73 & 9), drive northwest on Route 73 for 3.8 miles (or 2.4 miles from the bridge spanning the North Fork of the Boquet River) and park in a long pull-off on your right (44°08.220'N 73°44.548'W), 0.1 mile before the Zander Scott trailhead for Giant Mountain.

A short path leads directly to the waterfall's base, but you will see the cascade long before you reach it.

31. CHAPEL POND SLAB CASCADE & MINI TRAP DIKE

Diagonally across the road from Rock Garden Falls is an area that could rightfully be called the "Valley of the Giants"—a token nod to Giant Mtn., where everything is larger than life, including Roaring Brook Falls. Huge boulders, some the size of automobiles and buses, create a landscape of passageways and hideouts. One huge slab of rock has fallen in such a way as to create an enormous rock shelter. Many of these monoliths are used by rock climbers for bouldering.

Next to these boulders is Chapel Pond Slab, one of a series of giant rock slabs in Chapel Pond Pass that includes Emperor Slab and King Slab. The slab rises to a height of over 700 feet and is part of the northeast shoulder of Round Mtn. (~3,054'), whose heights rock climbers frequently aspire to scale.

Flowing down the rock slab is Chapel Pond Slab Cascade, a *very* seasonal cascade created by a tiny rivulet that follows a narrow, shallow channel down a bare rock-face incline for hundreds of feet. Give this tiny rivulet another couple of thousand years, and the channel containing it will undoubtedly be worn deeper, turning the shallow notch into a flume. Visit any time other than early spring, however, and you will probably see little to nothing for your efforts.

In *At the Mercy of the Mountains: True Stories of Survival and Tragedy in New York's Adirondacks*, Peter Bronski recounts a terrible accident that occurred at Chapel Pond Slab in the winter of 1975 when three ice-climbers fell from the heights, entangling a party of four climbers lower down on the slope.

Mini Trap Dike is a towering dike stacked with gigantic boulders, almost like a rock-filled chasm turned vertically on its end. Water comes cascading down through and over these boulders, producing miniature cascades, none of which are notable in themselves but which collectively produce an effect of water falling from a great height.

To get there: *To Chapel Pond Slab Cascade*—From the pull-off for Rock Garden Falls (see "Rock Garden Falls" for directions), walk across Route 73 and continue southeast along the road for over one hundred feet, being vigilant of cars. At the end of the guardrail, follow a trail southwest for several hundred feet that leads up to an enormous and nearly vertical rock wall (44°08.133'N 73°44.587'W). The field of giant boulders, slightly to the south, is impossible to miss along the way.

From Chapel Pond Slab Cascade to Mini Trap Dike—Head northwest along the rock wall for 100–150 feet. A climbing path, of sorts, follows up along the south side of the dike. This presumably is used by rock climbers and hikers to access higher regions.

32. BEEDE BROOK FALLS

Giant Mountain (4,627'), or "Giant of the Valley" as it was called during the nineteenth century, is the twelfth highest summit in the High Peaks Region of the Adirondacks. Without question, thanks to its proximity to Route 73, it is the most frequently summited among the tallest High Peaks. Giant Mountain was first climbed in 1797 by Charles Brodhead while in the area surveying the southern boundary of the Old Military Tract. His ascent was also the first one of an Adirondack peak over 4,000 feet in height.

One of Giant Mountain's distinguishing features is the Giants Washbowl, a small glacial pond that is seen by virtually everyone climbing up the mountain on the Zander Scott Trail.

Rarely visited, however, is Beede Brook, aka Dipper Brook and the Giants Dipper—a medium-sized stream that rises in part from Dipper Pond, aka Fingerbowl Pond, and another unconsolidated source from a col between Giant Mountain and Rocky Peak Ridge. Beede Brook flows into Chapel Pond after going under Route 73.

The reason for Dipper Pond and Beede Brook's obscurity is quite simple. There are no trails that lead to them. Access is only by bushwhack, and not an easy one at that.

I first did this hike twenty-five years ago, bushwhacking over from the Zander Scott Trail. I remember seeing a 10-foot-high waterfall followed by a 15-foot-high fall, both situated in a deep gorge with nearly vertical walls. Undoubtedly I must have passed by other cascades, but I have no recollection of them. What I do remember, however, is that it was a fairly demanding bushwhack.

In November 2016 I decided to approach the cascades in a different manner, this time by bushwhacking northeast up from Route 73 along the west bank of Beede Brook. Much to my astonishment, within 0.2 mile Beede Brook's streambed turned into a mammoth gorge with walls that quickly reached to nearly 100 feet in height. This was something that I had not expected. What's more, the gorge, with its perilously high walls, continued unabated for the rest of the trek up to the main falls.

Upper Beede Brook Falls is your reward for undertaking a demanding bushwhack.

As I scrambled along the steep slope above the canyon, I could occasionally hear the sound of rapids and small cascades below. Unfortunately, at no time was I was able to get close enough to look down into the gorge to see them. Still, just hearing them was sufficient inducement for me to keep pushing on.

At 0.4 mile I finally reached a spot where I could look down from above and glimpse two small-to-medium-sized cascades in the streambed below (44°08.220′N 73°44.115′W). However, there was simply no way to scamper down for a closer look, and I began to wonder if this was going to be an ongoing problem. I also began to question whether it would have been better if I had just stayed inside the gorge from the beginning, rock-hopping my way upstream. Still, I continued on.

After bushwhacking along the steep slope for another 0.2 mile (or well over 0.5 mile from the start), I again heard the sound of waterfalls. This time the canyon wall was less vertical and I was able to descend far enough into the gorge to catch a view of what I am calling Lower Beede Brook Falls—a 30-foot-high cascade enclosed in a narrow section of the gorge surrounded by soaring walls of immense height (44°08.317′N 73°43.986′W). At the top of the waterfall was a huge boulder that had become wedged in the channel. For the moment it seemed immune to the pressure being exerted on it by the onrushing stream. Someday, I knew, it would be dislodged and flung out onto the streambed below.

Regrettably, without ropes I was not able to reach the bottom of the gorge in order to take unobstructed photos of the waterfall. I ended up hovering around 15 feet above the streambed. Nevertheless, even partially blocked by tree limbs, I found the view of the waterfall to be astonishing—like something one would expect to see in Yosemite Valley.

Climbing back up, I continued along the top of the gorge for several hundred feet and then descended to the bottom of the canyon via a series of drops just upstream from the top of Lower Beede Brook Falls. At once I found myself in a huge gulf with towering rock walls all around me. Straight ahead was Beede Brook, filled to the brim with piles of large boulders that extended downstream to the top of Lower Beede Brook Falls. But most exciting of all was the partial sight of a large waterfall coming in on my left, nearly perpendicularly to the gorge. To get a head-on view I scrambled across the boulder-filled stream to the east side of the gorge and looked back. Before me was Upper Beede Brook Falls—an enormous, 80-foot-high cascade that dropped nearly vertically into a large pool of water created by the boulder field (44°08.305′N 73°43.940′W). It was simply dazzling. Adding doubly to the waterfall's stature was a huge cavern that had been worn into the bedrock along its southwest side. If you have a copy of the 1987 edition of Barbara McMartin's *Discover the Northeastern Adirondacks*, you can see a picture of the waterfall next to the title page.

Because of the lateness of the day, I had to cut short the length of my hike, not wanting to use up my margin of safety metered out by the amount of daylight remaining. I had learned the hard way that you don't want to be in the woods when night falls. Twenty-five years ago a friend and I had gone out to explore Mitchells Cave near Sprakers. We entered the cave in the early evening, only to emerge an hour or two later into total darkness — a situation

Multiple cascades of all sizes are encountered on Beede Brook.

that we had not anticipated. Even with our caving headlamps blazing, we were unable to follow the faint, unmarked, half-mile-long trail that had led us in. The woods seemed to have closed in around us and, for a moment, it looked like we were doomed to spend the night in the forest, much to the delight of the hundred thousand mosquitoes that had gathered around us to dine. But suddenly, deliverance came in a most unexpected way. I found that by listening carefully, I could hear the faint hum of cars moving along the New York State Thruway. At that instant I knew we could make our way out of the woods—and we did—all the way down to the Thruway, followed by a very, very long hike back to my car.

Without doubt the Beede Brook Falls are some of the most impressive waterfalls in the Adirondacks, made all the more spectacular by their remoteness. The hike to Beede Brook Falls, however, is one of the most challenging treks in this book and should only be attempted by a group of 2–3 hikers well equipped with a compass and topographical map if the gorge is being approached from the Zander Scott Trail. If you are approaching from Route 73, the task of navigating is slightly easier. You only need to follow the gorge up and then back down to avoid getting lost.

Getting there: From Underwood (junction of Routes 73 & 9), drive northwest on Route 73 for 3.8 miles (or 2.4 miles from the bridge spanning the North Fork of the Boquet River) and park in a long pull-off on your right just after crossing over Beede Brook (44°08.220′N 73°44.548′W). This is the same parking area as for Rock Garden Falls.

Option #1: Bushwhack uphill along rim of Beede Brook Gorge—Follow Beede Brook upstream along its west bank, staying several hundred feet away from the rim of the gorge. This part of the bushwhack is not very difficult unless you are scouting for waterfalls. In that case you will be scrambling along the steep slope near the rim, and the bushwhack becomes very strenuous indeed.

Regardless of your approach, the hardest part is getting down into the gorge to see the waterfalls close-up. This should only be attempted by hikers who are confident in being able to scramble up and down steep terrain and who are joined by two or three capable companions. Doing this hike solo is asking for trouble should you get hurt and become unable to exit under your own power.

Option #2: Bushwhack to falls from Zander Scott Trail—Start at the Zander Scott trailhead for Giant Mountain (44°08.309′N 73°44.621′W), 0.1 mile northwest of where Route 73 crosses Beede Brook. The trail, established in 1954, has become the most popular route for summiting Giant Mountain.

After following the trail uphill for ~0.4 mile to an altitude of ~2,150 feet (or ~500 feet above Route 73), head southeast through the woods to reach Beede Brook, which is less than 0.4 mile away. Upon reaching the gorge, you can now proceed either uphill or downhill, depending upon where you came in. Later, by following the gorge all the way back down to Route 73, you will end up less than 0.1 mile from the Zander Scott trailhead.

Take note that you can obtain a distant view of Dipper Pond by continuing up the Zander Scott Trail to an overlook ~0.3 mile above the junction with the Giants Nubble Trail.

33. CHAPEL POND FALLS

Chapel Pond Falls is a 50-foot-high, pencil-thin cascade formed on a tiny stream that drops into 19-acre, 0.5-mile-long Chapel Pond from the upper reaches of Round Mtn. The waterfall is located approximately at the point where Chapel Pond narrows and goes around a bend as it continues northwest for another 0.1 mile. In the early spring you can hear the fall faintly in the distance if you stand near the shore and listen closely. Unfortunately, most of the cascade can only be seen at times of heavy rainfall or snowmelt and under the right conditions of lighting. The bottom portion of the cascade generally remains hidden behind heavy growth.

What makes the cascade difficult to access directly is that it cannot be reached on foot. You must take along a canoe or kayak if you want to see it when water is flowing. There simply is no other way to get to it during most of the year unless you are a rock climber.

I have tried twice to get close to this cascade without recourse to rock climbing or paddling, exercising the full range of my creativity. In the first instance I waited until winter and walked 0.2 mile across frozen Chapel Pond to where I estimated the cascade would be. As it turned out, the entire rock face was streaked with frozen rivulets, making it impossible to know with certainty which one was the actual cascade and which were merely streaks of water that had frozen on the rock wall.

What I did observe were several parties of ice climbers making their way up the frozen slabs of ice at Chapel Pond despite the frigid temperature and gusty winds. Ice climbing has become a big winter sport in the Adirondacks, and the walls at Chapel Pond, as well as actual waterfalls such as Roaring Brook Falls, have become major attractions. For readers interested in pursuing this sport safely and intelligently, Don Mellor's second edition of *Blue Lines: An Adirondack Ice Climber's Guide* is hard to beat.

The second time, I hiked southeast over to a point of land where the west side of Chapel Pond begins to narrow. Even though only 0.05 mile of water separated me from the rock wall, I was still not able to see the waterfall because of its limited water flow and all the lush growth around its base.

Evidently, Chapel Pond Falls simply lacks the waterpower to wash away any debris or brush that has grown up or accumulated around its base, and so remains concealed.

This pencil-thin cascade at Chapel Pond rarely shows itself. Photograph by John Haywood.

Unlike me, be smart and take along a canoe or kayak to access this ephemeral cascade.

Even if you are not looking for waterfalls, a stop at Chapel Pond is well worth the investment of time. The pond is probably the most well-known frequently visited body of water in the heart of the Adirondacks. As James R. Burnside writes so poetically in *Exploring the Adirondack 46 High Peaks*, "The fluted columns of the cliffs on either side soar like organ pipes in a natural cathedral."

To get there: From Underwood (junction of Routes 73 & 9), proceed northwest on Route 73 for 4.1 miles (or over 2.7 miles from where Route 73 crosses the North Fork) and turn left into the parking area directly in front of Chapel Pond (44°08.435′N 73°44.799′W). Look for the waterfall on the opposite side of the pond, slightly to your right.

There is a second, less well known approach to Chapel Pond. From the parking area continue driving downhill on Route 73 for another 0.3 mile and then turn left onto a short dirt road that terminates near the northwest end of the pond. From there you can bushwhack southeast, following the shoreline, until you reach the point where the pond suddenly widens out (44°08.422'N 73°44.985'W). Perhaps you will have better luck glimpsing the waterfall from here than I had.

34. BEER WALLS FALLS

This tall, slender cascade is formed in an expansive gorge called the Chapel Pond Canyon, or Beer Walls. The canyon consists of a deep, mostly dry, rock-filled drainage area running northwest from Chapel Pond (but one that is not part of Chapel Pond's main outflow). As Elizabeth Jaffe and Howard Jaffe write in *Geology of the Adirondack High Peaks* (1986), "On the topographic map you will notice another stream to the west of this one, also parallel to the fault zone and separated from the main drainage by a narrow ridge. This stream does not quite tap Chapel Pond and is an abandoned channel, captured by the main drainage. Perhaps it was blocked by debris, or perhaps continued movement along the fault zone deepened the eastern channel." During high water, however, it may serve as an overflow channel.

The canyon was "discovered" sometime around 1982 by rock climber and author Don Mellor while in an airplane flying over the High Peaks scouting for potential rock climbing routes. Since then the canyon's northwest walls have become a popular rock climbing site, conveniently located near Route 73. The first climbing route established was called Positive Reinforcement. Others quickly followed.

Beer Walls Falls faces these cliffs from the opposite side of the canyon, which is a good thing, for it enables you to see the waterfall without having to first cross to the other side of the gorge.

Beer Walls Falls is a rather ethereal cascade that makes its appearance in the early spring, only to disappear in the summer if there is little rainfall. Altogether, including its upper, less visible sections, the waterfall drops over 150 feet.

Unlike other rocky bluffs in the Chapel Pond area that face the road, Beer Walls is separated from Route 73 by a forested ridge, creating the illusion once you are there that the canyon is far removed from civilization. Hiking in is like being transported into another world.

Beer Walls Falls is ethereal, much like the slender waterfall at the Cascade Lakes. Photograph by John Haywood.

It should be mentioned that the ledges and waterfall were not named to commemorate partying beer-drinking rock climbers (few rock climbers drink alcohol while preparing for or doing a climb, for obvious reasons—staying alive). The walls are named after a family of early settlers named Beers who lived in the valley. They also still live on in name at Beers Bridge Way, a dirt road that enters Route 73 between St. Huberts and Keene Valley near Mossy Cascade.

To get there: From Underwood (junction of Routes 73 & 9) drive northwest on Route 73 for 4.6 miles (or 0.5 mile northwest from the parking area at Chapel Pond). Heading downhill, turn into a pull-off on your left. If you are approaching from the opposite direction, drive 0.8 mile uphill on Route 73 from the Roaring Brook Falls parking area and turn right into a pull-off (44°08.710'N 73°45.274'W).

From the pull-off, walk ~200 feet down the road, going northwest. You will come to a tiny log bridge spanning a ditch where a little stream comes down through a small ravine. This is the trailhead. Follow the path uphill, going southwest. Within 0.1 mile you will reach the top of the ridge, where an outhouse can be seen. Bear right and follow the ridge-line path northwest, climbing slightly higher and eventually coming up to a large area of bare rock that overlooks the canyon. Expect to do a little bit of exploring to find the exact spot. You will know that you are in the right place when directly across the canyon is Beer Walls Falls. There are two other overlooks, but I would suggest heading to the westernmost one.

I have descended to the bottom of the canyon and walked across to the other side to see the cascade close-up. It was like walking across a dry rocky riverbed partially covered by moss and growth and filled with talus that had been brought down from the opposing walls of the chasm—very easy to twist an ankle. I wouldn't recommend doing it for the simple reason that the view of Beer Walls Falls didn't improve the closer I got to the waterfall.

You will find this a fascinating place to visit. Listen carefully and, if you're there at the right time of the day, you will likely hear the muffled voices of rock climbers echoing through the chasm as they call to one another.

35. ROARING BROOK FALLS

One of the most spectacular waterfalls in the Adirondacks is Roaring Brook Falls, contained in the 23,116-acre Giant Mountain Wilderness Area. It is a very high waterfall that is visible to your left while driving south toward the Adirondack Northway along Route 73. Approaching from the opposite direction, you will have to know when to look; otherwise, you will drive right past the view without realizing it.

According to *Heaven Up-h'isted-ness!: The History of the Adirondack Forty-Sixers and the High Peaks of the Adirondacks* (2011), "The falls are one of the Adirondacks' highest and the highest in the High Peaks. ... The current figure comes in at about 325 feet, only 25 feet lower than T-Lake Falls, considered the Adirondacks' tallest." This essentially matches what T. Morris Longstreth wrote nearly one hundred years ago in his 1922 book, *The Adirondacks*: "Roaring Brook Falls comes down three hundred feet, leap by leap in the beautiful woods of the Giant."

The waterfall is formed on Roaring Brook, a fairly modest-sized stream that rises from the west side of Giant Mountain near its summit and flows into the East Branch of the Ausable River. In *By Foot in the Adirondacks* (1972), Phil Gallos writes, "Roaring Brook Falls is divided into three sections. First there is a long vertical plunge, then a short cascade, and, finally, another long vertical."

Like any stream fueled by a high mountain, Roaring Brook can turn into a monster when it becomes over-energized by rainfall or sudden releases of snowmelt. Indeed, such an event occurred on June 29, 1963. So much rain fell within a short period of time that a huge avalanche of mud, debris, and water broke loose from the heights. It crashed down over the waterfall and raced through the valley below, sweeping everything in front of it and leaving behind a 150-foot-wide swath. The deluge was so powerful that it swept right across Route 73, literally a river of mud many feet high. Nearly 400 yards of highway were covered. Cars were mired in the mud and debris, and the road was closed until crews were able to clear away the sludge.

The sudden deluge and landslide actually altered the path of the stream, diverting it north into Putnam Brook, a route that reportedly the stream is said to have taken after a similar event occurred in 1856. A crew of four men worked for 10 days, erecting crib dams where necessary near the top of the fall, in order to restore the river back to its present course. Today, Roaring

Brook Falls races down through a massive area of exposed anorthosite bedrock. It will take hundreds of years for the invasive power of life to cling again to the rocks and turn the waterfall into a structure of both earth and water.

Except for T-Lake Falls, Roaring Brook Falls is the highest waterfall in the Adirondacks.

In his 2008 book, *In Stoddard's Footsteps*, Mark Bowie presents a circa 1880s photograph by Seneca Ray Stoddard and a 2000s photo of his own, both taken of Roaring Brook Falls from the same vantage point. It's a fascinating way of depicting the kind of changes that have occurred over a period of 125 years. A photograph of the waterfall also appears in James Kraus's *Adirondack Moments* (2009), and Hardie Truesdale's *Adirondack High: Images of America's First Wilderness* (2005).

In *Rocks and Routes of the North Country New York* (1976), Bradford B. Van Diver states that four conditions must be present in order for a landslide of the magnitude of the 1963 slide to occur. First you need a bowl-shaped cirque where huge volumes of water are rapidly funneled into a narrow channel. Second, the soil on top of the bedrock must be thin and ready to yield if given a sufficient push. Third, the underlying bedrock has to be smooth from glacial scouring so as to offer little resistance to the soil sliding over it. And fourth, the bedrock must be solid enough without joints, so that trees, which would otherwise block the landslide, are easily uprooted and swept away.

As with any large waterfall, deaths are associated with Roaring Brook Falls. People have fallen off the top and died, something that is not as hard to do as you might first think. The bedrock next to the fall's summit is curved and smooth. Should you step out too close to the rounded edge when the rocky surface is wet, slick, or icy, well—you can imagine the rest. There is no way to ride this waterfall down without being reduced to a heap of broken bones at the bottom. Peter Bronski in his 2008 book *At the Mercy of the Mountains* describes the perils of Roaring Brook Falls only too well: "… from this expansive vantage point, the water flows over a concave lip of rock. Careless hikers wanting a glimpse down the falls inch forward, lured by the slowly developing view, until the friction of their sneakers or boots can no longer withstand the pull of gravity on the steepening rock."

As recently as June 2016, a thirty-seven-year-old Staten Island woman was found dead at the base of the waterfall. I would hazard a guess that at least a dozen people have died at the waterfall over the years.

A twelve-year-old Greenwich boy was killed at the fall in March 2016, but not from falling. He was standing at the base when he was struck by a 600–800-pound boulder that fell from the heights. Such a freak accident is a sober reminder that anything can happen at a large waterfall. Stay vigilant—always.

In *The Adirondacks Illustrated*, Seneca Ray Stoddard quotes Old Mountain Phelps, who had this to say about the waterfall: "See that bare rock near Smith Beede's? These are Roaring Brook Falls, the highest in the mountains; nearly 200 feet sheer fall at one leap, and I tell you it isn't much besides spray when it reaches the bottom." Smith Beede's place presumably was by Putnam Brook, part of which runs parallel to Roaring Brook.

Ironically, the best view of Roaring Brook Falls is from Route 73, where you get to see the "big picture." There is a roadside pull-off that allows for lingering views and photos. Counter-intuitively, the closer you get to Roaring Brook Falls the less you see, thanks to the waterfall's architecture and enormous height. Standing at the base of the fall and looking up reveals only its lower section.

The views from the top of the waterfall are extraordinary and quite expansive—not so much of the waterfall, but rather of the Great Range to the southwest and Noonmark Mtn. (3,556') to the south. If you have a topographical map in hand, you can count off the names of the mountains in the distance.

During the winter the waterfall is frequented by ice climbers. Because the stream falls like a sheet over the bedrock, it freezes fairly quickly and can be relied upon for solid holds when engaged by a handheld ice ax. A photo of ice climbers ascending the waterfall in a column can be seen in Mark Bowie's 2006 book, *Adirondack Waters: Spirit of the Mountains*.

To get there: From Underwood (junction of Route 73 & 9) drive northwest on Route 73 for 5.2 miles to a roadside pull-off to your right with superior views of Roaring Brook Falls in the distance (44°08.951'N 73°45.787'W); or drive another 0.3 mile (a total of 5.5 miles) to the parking area on your right for Roaring Brook Falls and Giant Mountain (44°09.025'N 73°46.050'W). The trail up past Roaring Brook Falls to the summit of Giant Mountain was cut in 1874 by Orlando Beede and was used as a bridle path for horses and riders, getting them up to nearly 300 feet from the summit.

To base of fall—From the parking area follow the main trail east for 0.1 mile and then continue on a well-worn path across the valley floor, continuing east, for another 0.2–0.3 mile to the base of Roaring Brook Falls.

To top of fall—From the parking area follow the main trail east for 0.1 mile. At a junction bear left, continuing on the main trail as it heads steadily uphill, initially paralleling the spur path in the valley below. Not far from the top of Roaring Brook Falls, look for a massive glacial boulder to your left about 50 feet from the trail.

After ascending 0.4 mile from the valley, bear right at a junction and walk 150 feet south to reach the top of Roaring Brook Falls.

IV: ADIRONDACK MOUNTAIN RESERVE

The Ausable Club.

The East Branch of the Ausable River rises from Lower Ausable Lake and joins with the West Branch of the Ausable River at Ausable Forks to form the main trunk of the Ausable River. The Ausable is a formidable river and one that can easily become uncontrollable. In the September freshet of 1856, the 10-foot-high Wells Dam at Lower Ausable Lake gave way, sending a swell of roiling, thundering water downriver that annihilated nearly every mill and bridge from St. Huberts to Keeseville before dissipating into the deep waters of Lake Champlain. Lives were lost as well. A full account of this terrible incident was written in an article called "Violence in the Valley" by Laura Viscome in *Adirondack Life*'s *Adirondack Waterways: 2001 Collectors Issue*.

Since then similar episodes have occurred throughout the years. Starting in 1925 there have been over eleven "major flood stage" events on the East Branch (defined as high-water gauge readings exceeding 11 feet). The after-effects of recent Tropical Storms Sandy and Irene are all too evident reminders of the damage caused by flooding.

The waterfalls in this section are located on property owned by the Adirondack Mountain Reserve (AMR), accessible from trails off the Lake Road. The Lake Road—a 3.5-mile-long private road that was initially a toll road—was constructed between the Ausable Club and Lower Ausable Lake in 1892, the same year that the Adirondack Park was created. The road gains 700 vertical feet over the course of its 3.5 miles. It essentially parallels the East Branch of the Ausable River and is loosely classified as a trail since no vehicles are allowed on it except for a bus that is operated by the Ausable Club for its members and guests. Access by the public to all the trails in the Adirondack Mountain Reserve (with the exception of certain trails near the shore of Upper Ausable Lake) was made possible by a permanent public easement for foot travel obtained by the State of New York in 1978 as part of the Adirondack Mountain Reserve's sale of higher land.

The Adirondack Mountain Reserve was formed in 1887 and was principally spearheaded by William G. Neilson. Its formation involved the purchase of a huge tract of land, including the Upper and Lower Ausable Lakes and all or part of thirteen of the High Peaks. In 1897 the Adirondack Trail Improvement Society (ATIS) was founded by the Adirondack Mountain Reserve to help develop and maintain many of the Adirondack trails leading into the High Peaks. It is through their diligent efforts that the trails are so well maintained.

In 1890 members of the Adirondack Mountain Reserve formed the Keene Heights Hotel Company and erected the St. Huberts Inn on the site of the former Beede House (a hotel named after Smith Beede), which had just burned down. A photo of the Beede House taken sometime after 1876 can be seen in the 1999 book, *Two Adirondack Hamlets in History: Keene and Keene Valley*.

Ultimately the hotel venture failed. In 1904 the members of the Adirondack Mountain Reserve turned the hotel into an exclusive club, renaming it the Ausable Club in 1906. It remains a seasonal, private, 450-plus-member club today.

This section of the book covers the East Branch of the Ausable River from the Ausable Club to Lower Ausable Lake. To the west of the East Branch is the 10-mile-long Great Range. The mountain range consists of Snow Mtn. (2,375'), Hedgehog Mtn. (3,389'), Lower Wolfjaw Mtn. (4,175'), Upper Wolfjaw Mtn. (4,185'), Armstrong Mtn. (4,400'), Gothics (4,736'), Saddleback Mtn. (4,515'), Basin Mtn. (4,827'), Little Haystack (4,700') and, northwest, Mount Haystack (4,960'). To the southeast, on the opposite side of the East Branch, are the lesser, but still quite high, mountains of the Colvin Range, including Noonmark Mtn. (3,556'), Dial Mtn. (4,020'), Nippletop Mtn. (4,620'), Mount Colvin (4,057'), Blake Peak (3,960'), and Pinnacle Mtn. (3,349').

Between the Great Range and the Colvin Range, glaciers and the East Branch have cut a deep valley along a diagonal fault line running northeast-southwest, its waters fed by watersheds created by large mountains. Once the East Branch reaches Keene Valley, the river levels off, broadening, until it races over Hulls Falls and then through a narrow, treacherous gorge near Keene. From Keene the East Branch flows north through Jay and Upper Jay without further fanfare, passing through a broad river valley. Coming to Ausable Forks, the East Branch combines with the West Branch to form the Ausable River, flowing along fairly placidly until reaching Keeseville and its waterfalls. As Winslow C. Watson writes in *Military and Civil History of the County of Essex, New York* (1869), "Nearly the whole course of the Au Sable and its branches presents a series of falls, cascades, and rapids, which, whilst they adorn and animate the scenery, afforded innumerable sites of water power, rarely exceeded in capacity and position."

Lower Ausable Lake is about two miles long, covering 320 acres. Its shape is similar to a fjord—narrow, with steep precipitous walls. The lake is surprisingly shallow, with a maximum depth of 21 feet. Upper Ausable Lake, farther south, is slightly larger at 360 acres. The lakes came into existence when waters from ancestral Lake Keene were ultimately bled off into the Champlain Valley, leaving behind a much smaller body of water. Geologists believe that the two lakes were originally joined together until they became separated by alluvial deposits left behind by the inrushing waters of Shanty Brook. This isthmus created a natural dam between the two lakes.

In 1854 David Heald (who later changed his last name to "Hale") constructed a sawmill and dam at Lower Ausable Lake's outlet for Sylvanus Wells, a State Canal Commissioner and lumberman. The first dam was fairly primitive and was washed away in 1855. A second dam was built in 1856, raising the lake level by 15 feet. It was promptly washed away that same year by what is referred to as the "freshet of 1856." In 1857 a third dam was erected, this time increasing the lake's length by 0.5 mile. Log drives using controlled dam releases of water continued down the East Branch through the early 1920s until trucking provided a more expedient method of transportation. The most recent dam at Lower Ausable Lake was constructed in 2003–2004.

During the mid-to-latter nineteenth century, the natural beauty of the St. Hubert's area proved irresistible to artists. Hundreds were inspired by the breathtaking wilderness scenery, including such luminaries as Asher B. Durand, John William Casilear, John F. Kensett, Frederick Stanton Perkins, Roswell Morse Shurtleff, A.H. Wyant, Winslow Homer, Sanford Robinson Gifford, John Lee Fitch, John Adams Parker, Samuel Coleman, and Arthur Parton. It is said that the artist who started it all in 1857 was T.S. Perkins. A photo of some of the artists posed in front of the Ausable River can be seen in *Two Adirondack Hamlets in History: Keene and Keene Valley*.

It would seem that the majority of the hikers coming into this part of the Adirondacks are heading up to the mountains. It's a long-established custom. Many are looking to become 46ers, a tradition that goes back to the formation of the Adirondack Forty-sixers when two brothers, Robert and George Marshall, with the aid of their guide, Herbert Clark, finished climbing all forty-six peaks in 1925. Believe it or not, we even know the name of the first dog to become a Forty-sixer—Chrissie, in 1948.

I am pleased to be able to say that I was one of the first to write about the Adirondack Mountain Reserve from the perspective of its waterfalls. In 1998 an article of mine called "Hiking the Waterfall Trail" appeared in the May/June issue of *Adirondac*, in which I described many of the waterfalls that have formed on tributaries of the East Branch. Even at that time, without being fully conscious of it, I was trying to encourage hikers to make waterfalls a destination in their own right. The article eventually became the nucleus for my first waterfall guidebook, the *Adirondack Waterfall Guide,* published by Black Dome Press in 2003.

To get there: From Underwood (junction of Routes 73 & 9) drive northwest on Route 73 for ~5.5 miles and turn left onto Ausable Road. Park immediately in a large area to your left (44°08.981'N 73°46.067'W). Be aware that parking is not allowed anywhere along Ausable Road.

From the parking area, which provides access not only to the Adirondack Mountain Reserve trails but to Noonmark & Round Mtn. as well, walk south, proceeding uphill for 0.6 mile along a partially dirt road to reach the Ausable Club. In doing so you will pass by the club's 9-hole golf course (which was expanded to its present size in 1928) to your right. When you reach the tennis courts, turn left and follow the road southwest for 0.2 mile until you come to the gatehouse. Straight ahead is the 3.5-mile-long Lake Road, listed in the ADK guidebook as the Lake Road Trail.

Be sure to sign the trail register at the gatehouse and review all of the posted AMR rules before beginning your hike. These rules include no dogs, no cell phone use, no swimming, fishing, or boating at the Ausable Lakes, no crossing the lakes when frozen, no bicycling, and no bushwhacking—hikers must stay on marked trails.

FALLS ALONG THE WEST RIVER TRAIL

The West River Trail follows along the west side of the Ausable River's East Branch from southwest of the Ausable Club to Lower Ausable Lake. It provides the only practical means by which to access the large waterfalls coming down on tributaries to the East Branch from the west unless you were to climb over mountains along the Great Range and come down into the valley repeatedly.

A rugged section of the West River Trail.

To get there: At the gatehouse turn right where a sign indicates "Trail to Wolf Jaws via W.A. White Trail. West River Trail. Cathedral Rocks" and follow the trail northwest. As soon as you cross over the East Branch of the Ausable River via a footbridge rebuilt in 2012, turn left onto the West River Trail (44°09.034'N 73°47.008'W). At this point you will have gone nearly 0.2 mile. The yellow-blazed West River Trail parallels the East Branch all the way up to Lower Ausable Lake. It was first blazed in 1899. The William A. White Trail, which you are not taking, leads straight ahead up to Lower Wolf Jaw and the Great Range.

36. PYRAMID FALLS

Pyramid Falls is a 20-foot-high cascade formed on Pyramid Brook—a small stream that rises from a col between Hedgehog Mtn. (3,389') and Lower Wolfjaw Mtn. (4,175'). Although it would seem logical to assume that the source of Pyramid Brook must be 4,515-foot-high Pyramid Mountain, a sub-peak of Gothics, the two are actually separated by several miles and no connection exists between them. The name of the stream and waterfall likely came from the pyramidal shape of the cascade as its falling waters fan out during its descent.

A photograph of Pyramid Falls, listed as "Bear Run on the Ausable's East Branch," is shown in Eliot Porter's 1966 book, *Forever Wild: The Adirondacks*.

A 10-foot-high elongated cascade can be seen 50 feet downstream from Pyramid Falls where Pyramid Brook drops through a narrow flume.

To Pyramid Falls: Proceed south along the West River Trail, paralleling the East Branch, for 0.9 mile (or 1.1 miles from the start of the trail at the gatehouse). When you come to the red-blazed Pyramid Brook Trail (44°08.506'N 73°47.987'W), turn right and head uphill for less than 0.3 mile. As you cross over Pyramid Brook via a rock hop, look to your left, upstream, to see a 10-foot-high cascade dropping through a narrow flume (44°08.699'N 73°47.958'W).

On top of Pyramid Falls.

From the opposite side of the brook, continue uphill for a couple hundred feet more. Then take a short spur path that leads immediately to the base of Pyramid Falls where the stream is momentarily swallowed up by rocks (44°08.704′N 73°47.984′W).

37. FIRST FALLS ON EAST BRANCH OF AUSABLE RIVER

Within another 0.4 mile (or ~1.5 miles from the gatehouse), you will pass by an overlook with views into the interior of the East Branch Gorge where cascades can be glimpsed when there are few leaves on the trees to obscure the view. Look for a wooden railing (if it is still there). There is really no way to descend to the streambed for a closer look. The cascades are far below and the slope is exceedingly steep. A photo of this spot can be seen in James R. Burnside's 1996 book, *Exploring the 46 Adirondack High Peaks*. This is just one of several pretty cascades along the Adirondack Mountain Reserve's section of the East Branch that are better seen from the opposite side of the river (see "East River Trail").

38. WEDGE BROOK FALLS

Wedge Brook Falls consist of three separate, distinctly different waterfalls that have formed on Wedge Brook—a medium-sized stream that rises from the southeast shoulder of Lower Wolfjaw Mtn.

The upper fall (44°08.166′N 73°48.646′W) is ~75 feet high and located several hundred feet upstream (the sign says 200 yards). It can be reached from the Lower Wolf Jaw Mtn. Trail as soon as you cross over the footbridge spanning Wedge Brook.

The middle, 10–15-foot-high waterfall comes down in front of the footbridge spanning Wedge Brook (44°08.157′N 73°48.643′W). The stream and fall received its name from this tiny flume and its wedged rock.

The 25-foot-high lower fall is downstream from the footbridge and, while not as high as the upper fall or as flume-like in appearance as the middle fall, it provides the perfect setting for lunch or a respite while sitting on the bedrock along the stream. *In New York Waterfalls* (2010) Scott E. Brown describes it as "a three-tiered cascade that's emerald green and quite lovely."

Lower Wedge Brook Falls. Photograph by John Haywood.

In all likelihood most people trekking along the West River Trail glimpse the lower cascade and then stop for a moment on the footbridge to view the middle cascade. After that they are gone, thinking that the next big waterfall ahead is Beaver Meadow Falls. In doing so they will have missed Wedge Brook's biggest waterfall, just a short distance upstream from the footbridge.

To the footbridge crossing at Wedge Brook Falls: From the overlook at the First Fall on the East Branch, continue south on the West River Trail for another 0.5 mile (or a total of 2.0 miles from the gatehouse). You will come to Wedge Brook, where a pretty footbridge spans the stream (44°08.157'N 73°48.643'W). Be sure to take time to see all three of the waterfalls. It is rare to have so many uniquely different waterfalls in such close proximity to one another.

39. SECOND FALLS ON EAST BRANCH OF AUSABLE RIVER

This 15-foot-high flume cascade (44°07.975'N 73°48.680'W) is formed where the stream is momentarily compressed by side walls. Racing ahead 70 feet, the river then enters a shallower flume where it drops over a 4-foot-high cascade.

To Second Falls on East Branch: From Wedge Brook Falls continue south for another ~0.3 mile (or 2.3 miles from the gatehouse). The waterfall can be glimpsed below in the gorge from where the second of two tiny tributaries crosses the trail. I scampered down to the streambed to take some close-up photos, but the view from the trail is more than adequate.

40. BEAVER MEADOW FALLS

Beaver Meadow Falls is assuredly one of the most photographed waterfalls in the Adirondacks. It looms at an impressive height of 60 feet and cascades down a series of steps so narrow that the waterfall's appearance of near verticality is never compromised. Many have described the falls as being bridal veil in appearance. Elizabeth Jaffe and Howard Jaffe, in *Geology of the Adirondack High Peaks* (1986), describe the waterfall as "where the brook leaps over a steep embankment caused by a fault zone."

At one time a wooden bridge crossed directly in front of the waterfall, but it was washed away some years ago during one of Beaver Meadow Brook's rampages.

Beaver Meadow Falls is formed on Beaver Meadow Brook—a medium-sized stream that rises from the east and south slopes of Upper Wolfjaw Mtn. Its name comes from the meadow that Beaver Meadow Brook passes through as it makes its way downstream from the falls into the East Branch. The meadow is a very scenic area to photograph.

Many photos have been taken of Beaver Meadow Falls. Several of the prettiest are in Nathan Farb's 1985 classic, *The Adirondacks*, Den Linnehan's

2004 book, *Adirondack Splendor*, Eliot Porter's *Forever Wild: The Adirondacks* (1966), Carl Heilman II's *Adirondacks: Views of an American Wilderness* (1999), and Derek Doeffinger & Keith Boas's *Waterfalls of the Adirondacks and Catskills* (2000). There is also a large close-up black & white photograph of the waterfall in Howard Kirschenbaum's 1983 book, *The Adirondack Guide*.

Beaver Meadow Falls is very photogenic.

In her 1987 book, *Up the Lake Road: The First Hundred Years of the Adirondack Mountain Reserve*, Edith Pilcher mentions another waterfall on Beaver Meadow Brook higher up along the trail to Gothics. Apparently it was favored by earlier generations but has fallen into obscurity today, partially concealed behind brush and vegetation. An accompanying photograph in her book shows a cascade approximately 8–10 feet high dropping down a rocky streambed.

During times of heavy water flow, look for a medium-sized cascade that forms a hundred feet to the left (south) of Beaver Meadow Falls, virtually next to the enormous wooden ladder that takes hikers up to the Lost Lookout, Gothics, and Armstrong Mtn.

To Beaver Meadow Falls: Continue south from the Second Falls on East Branch of Ausable River, following the West River Trail for another 0.4 mile (or ~2.7 mile from the West River Trailhead) to reach Beaver Meadow Falls (44°07.772′N 73°48.992′W), located 100 feet upstream from a small footbridge spanning Beaver Meadow Brook.

For those approaching from the East River Trail, Beaver Meadow Falls is only ~200 feet up from the west end of the Beaver Meadow Falls footbridge. This bridge spanning the East Branch was rebuilt in 1999 in memory of George H. Bright. It is one of several bridges spanning the East Branch that have had to be rebuilt a number of times because of the river's destructive power and ferocity. The East Branch is not a river to be trifled with or underestimated!

41. RAINBOW FALLS

Rainbow Falls is formed on Cascade Brook, aka Rainbow Creek—a medium-sized stream that rises from a cirque between Gothics and Armstrong Mtn. There is nothing to prepare you for the size of this waterfall. It is a monster, topping out at a height of 150 feet.

In his 1869 *Military and Civil History of the County of Essex, New York*, Winslow C. Watson writes, "The fall is computed from careful observation to be one hundred and twenty-five feet in sheer vertical descent." Watson's estimate, as it turns out, is not too far off the mark from the one conventionally used today. Although not a woodsman by profession, Watson, an attorney, had a keen mind for making discerning observations.

In *The Adirondacks Illustrated*, Stoddard writes: "We crossed the outlet and went up into the cleft mountain side, very like Ausable Chasm and probably with a like origin. It extends only a short distance but is very beautiful, the gray sides perpendicular for something over a hundred feet, while huge rough boulders fill the bottom, and over the edge of the wall at the north is the Fall, a skein of amber silk that flutters along down the rocks until whipped and ravelled [sic], it reaches the bottom as lightly as a snowflake falls and white as clean wool, gathering its drops together, it goes softly singing down its emerald-paved steps to the river below." In all likelihood Stoddard probably visited the waterfall during the summer when there was little water to animate Cascade Creek.

150-foot-high Rainbow Falls towers over a massive, boulder-filled chasm. Photograph by John Haywood.

James R. Burnside, in *Exploring the 46 Adirondack High Peaks*, offers a contemporary description of the waterfall, which "even in midsummer dry periods spills shimmering sheets to a pool 150 feet below. The flow begins over an overhanging rock so that on occasion the space behind creates a prism of colors in the morning sunlight."

In his 1976 book, *The Adirondacks*, Clyde H. Smith points out a fact that is relevant if you are visiting the waterfall early in the season. "In spring, vapor generated by its gigantic thundering pours from the grotto like a small rainstorm. On a few occasions when I tried to hike up the narrow ravine that cradles the cataract, I became drenched long before actually seeing the falls." Smith goes on to say, "Later, in summer, when its volume diminishes, Rainbow Falls seemingly flows in slow motion, releasing lacy veils that vaporize and disappear."

In their 1986 book, *Geology of the Adirondack High Peaks*, Elizabeth Jaffe and Howard Jaffe write, "Both the dike at the base of the falls—which is well exposed under water—and the anorthosite are considerably sheared as the Ausable Lakes form part of an extensive fault zone."

Rainbow Falls in the early spring.

What makes Rainbow Falls so appealing are both its remoteness and the depth of the canyon encasing it. Everything is super-sized here. The waterfall tumbles off the top of a sheer rock wall to crash onto the rocks and boulders far below. The hike into the canyon is particularly thrilling, requiring that you scamper around and over large boulders.

Rainbow Falls was first scaled in the winter of 1952 by David Bernays, who managed the climb by cutting steps into the center of the ice wall. At some later date the Ausable Club put an end to this activity, perhaps for aesthetic reasons or concerns about liability. Even so, deaths are associated with the waterfall. In 1993 a rock-climber fell to his death after a hand-hold broke loose.

Rainbow Falls, like Buttermilk Falls and High Falls, is a name that has been applied repeatedly to large waterfalls in New York State. There are 70-foot-high Rainbow Falls in Ausable Chasm (see "Ausable Chasm"); 20-foot-high Rainbow Falls at High Falls Gorge in Wilmington Notch (see "High Falls Gorge"); 10-foot-high Rainbow Falls on the Middle Branch of the

Oswegatchie River; 35-foot-high Rainbow Falls on the South Branch of the Grasse River; 90-foot-high Rainbow Falls in Watkins Glen; and 50-foot-high Rainbow Falls in Minnewaska State Park Preserve in the Shawangunks. I suppose that any waterfall that throws up a prismatic spray to produce a rainbow can justifiably be called "Rainbow Falls." One must wonder then why Niagara Falls, with its huge rainbow, *wasn't* called Rainbow Falls when visitors first encountered it.

According to Jerome Wyckoff in his 1979 book, *The Adirondack Landscape*, "Rainbow Falls is one of many Adirondack waterfalls that has formed in a dike where a wide, wall-like mass of less resistant rock that once intruded into a pre-existing fractured section of bedrock was eventually eroded away by stream action. "

Photographs of Rainbow Falls can be seen in Nathan Farb's *Adirondack: Wilderness*, Den Linnehan's *Adirondack Splendor* (2004), Den Linnehan's *New York State Splendor* (2008), Edward M. Smathers, Scott A. Ensminger, & David J. Schryver's *Waterfalls of New York State* (2012), Clyde Smith's *The Adirondacks* (1976), and Gary A. Randorf's *The Adirondacks: Wild Island of Hope* (2002).

To Rainbow Falls: From Beaver Meadow Falls, continue south on the West River Trail to its end by Lower Ausable Lake, a hike of less than 0.9 mile (or over 4.0 miles from the gatehouse). Along the way the trail leads away from the river momentarily, taking you through an area where the path is not as heavily worn, sometimes marshy, and therefore slightly harder to follow. You will also pass along the base of an enormous rock wall. When you finally reach the dam at Lower Ausable Lake where a long footbridge crosses over to the east bank of the Ausable's East Branch, cross over a smaller footbridge that spans Cascade Brook (the stream on which Rainbow Falls has formed) and turn right. Follow the "Gothics. Sawteeth. Rainbow Falls" trail west for 0.1 mile. Then bear right at a junction, taking a spur path marked by a sign indicating Rainbow Falls. This leads in less than 0.1 mile to the interior of a huge box canyon containing Rainbow Falls (44°07.102'N 73°49.768'W). If you are visiting in the early spring, be prepared for ice and snow that linger for months in the deep recesses of the chasm.

Rainbow Falls can also be viewed from the south end of the Lost Lookout Trail, less than 0.2 mile uphill from the West River Trail. A photograph of the fall from this vantage point can be seen in Tim Starmer's 2010 book, *Five-Star Trails in the Adirondacks*.

Shanty Brook Falls—There is a waterfall farther southeast of Rainbow Falls above Upper Ausable Lake, but it is difficult to get to and requires summiting

Sawteeth (4,100') from Lower Ausable Lake and then descending virtually all the way down to the shore of Upper Ausable Lake (and then repeating the hike in reverse to return to the dam by Lower Ausable Lake).

I am calling the waterfall "Shanty Brook Falls" since it is formed on Shanty Brook—a small stream that rises from cols between Basin Mtn. (4,827'), Saddleback Mtn. (3,515'), Gothics (4,736'), and Pyramid Peak (4,550'). Its GPS reading is approximately 44°05.180'N 73°51.955'W. Undoubtedly there are additional cascades farther up the stream that would require a bushwhack to access, but bushwhacking is not allowed on AMR property. The fall is enticingly showcased in Edith Pilcher's *Up the Lake Road* (1987) when she writes: "Shanty Brook, which crosses the trail, has always been a favorite picnic and fishing site. Youngsters enjoy sliding down the two-step water slide, while their elders simply feast their eyes and ears upon the beauty of the waterfall." Obviously, Shanty Brook Falls is not difficult for members of the Ausable Club to access, since they can paddle from Lower Ausable Lake to the Warden's Camp at the north end of Upper Ausable Lake and end up only 0.2 mile from the waterfall.

Lower Ausable Lake Dam.

42. FALLS ALONG THE EAST RIVER TRAIL

The East River Trail follows along the Ausable River's East Branch between Lower Ausable Lake and the Ausable Club. A number of cascades can be seen along the trail, several of which can only be partially glimpsed from the West River Trail. My recommendation is that you hike in on the West River Trail and hike back out on the East River Trail. That way you get to experience the river and all of its features from different vantages without duplication.

A long footbridge spans the East Branch of the Ausable River just downstream from the Lower Ausable Lake Dam.

To get to the East River Trail: From the end of the West River Trail at Lower Ausable Lake, walk across the footbridge spanning the East Branch downstream from the dam at the north end of Lower Ausable Lake, turn left at the end of the footbridge (44°07.112'N 73°49.490'W), and then begin hiking north on the East River Trail.

    The first part of the trek is fairly uneventful. The only thing you will pass by of note is the Bullock Dam, which consists of a wooden platform on top of a long log spanning the river. The cascades begin after the Beaver Meadow Bridge Trail is crossed.

Continuing past the intersection with the connecting trail to Beaver Meadow Falls, you will come to a high overlook of the East Branch within 0.5 mile where the river below drops violently, producing a tumultuous cascade. The GPS reading here is 44°07.996′N 73°48.674′W. A couple of hundred feet farther downriver is another high overlook with spectacular views of a second cascade. During times of high water flow in the early spring, 20–25-foot-high cascades are produced on two tributaries dropping into the gorge here. Look off in the distance above the second tributary-produced cascade and you will see an enormous waterfall. What you are looking at, in fact, are the three Wedge Brook Falls, seemingly stacked up on top of one another to form one continuous cascade.

Wild, dynamic views of the East Branch continue unabated along the East River Trail. You will never be far from the river, although frequently high above it. This is not a path that faint-hearted hikers should travel.

For those wishing to return to the parking area more quickly and more expediently than by hiking the East River Trail, there is the option of walking back along the Lake Road, which is easier and shorter, but hardly as scenic.

43. RUSSELL FALLS

Russell Falls is formed on the East Branch of the Ausable River downstream from the West River Trailhead. It is on private land, hidden away in the woods behind the Ausable Club and not accessible to the public. The waterfall is block-shaped with a small channel near its middle. It is the last major waterfall on the East Branch until Hulls Falls in Keene is reached (see "Hulls Falls"). Topographical software gives Russell Falls a GPS coordinate of around 44°09.159′N 73°46.967′W.

Some readers may wonder if I made up this name for reasons of vanity, my middle name being "Russell." Not so. Look at a Delorme *New York State Atlas & Gazetteer* and you will see the name Russell Falls clearly listed on the map. My guess is that Russell Falls was named for an early settler, landowner, or member of the Ausable Club. The name clearly predates 1874, at which time Seneca Ray Stoddard in *The Adirondacks Illustrated* quotes Old Mountain Phelps as saying: "You see that chasm there? That is the lower end of Russell Falls. There is a gorge through that hill near 200 feet deep, the width of the river, and nearly perpendicular walls on either side, a continuous ragged fall all the way for half a mile, and at no place more than 25 feet at one leap, but there is a great variety in them."

There is also a Russell Falls in the Central Highlands Region of Tasmania, Australia, but I assure you I did not name that one either.

FALLS ON GILL BROOK

Gill Brook rises from Elk Pass between Nippletop and Mount Colvin and flows into the East Branch of the Ausable River roughly 0.8 mile upstream from the Ausable Club. A large number of cascades have formed on this stream, the most famous being Artist Falls, named for its close association with nineteenth-century artists who came up to Keene Valley and St. Huberts to paint and write about what they saw.

That artists favored the falls on Gill Brook over the myriad of falls along the west side of the East Branch was due to one simple fact—Gill Brook was easy to get to. In contrast the falls on the west side of the river required that one ford the raging East Branch to reach them, and that was not something one easily did with easel and palette in hand. Gill Brook also offered to artists a series of waterfalls on the same stream, as opposed to those on the west side of the river that were contained on different streams separated by considerable distances.

How Gill Brook acquired its name is unknown, at least to me. Most likely it came about from someone named Gill who was associated with the brook.

An 1874 painting by Alexander Lawrie called *Gill Brooke* can be seen in *Two Adirondack Hamlets in History: Keene and Keene Valley* (1999). Samuel Colman and James David Smillie—two nineteenth-century painters—were also particularly drawn to paint Gill Brook and its falls.

In his 2004 book, *Adirondack Splendor*, Den Linnehan presents a series of images capturing some of the Gill Brook waterfalls. These are all spectacular waterfalls.

One of the hikes we led to Gill Brook will always stick in my mind. Barbara was laid up back at the inn with back problems, so it was just me out there with 13 hikers. Lucky number, right? Worst of all, the rain had started and was now coming down like it did when Noah launched his ark. As we headed up Ausable Road to begin the hike, I kept thinking that it's got to let up, but it never did. When we finally reached Gill Brook, the waterfalls were fantastic, but it was raining too hard to take pictures. The goal at this point shifted from hiking farther up Gill Brook, perhaps to as high as Fairy Ladder

Falls, and changed instead to going up to the top of Indian Head, down to Lower Ausable Lake, and then out via the Lake Road Trail.

The trail up to Indian Brook from Gill Brook is very steep—*very* steep. Unknown to me, the deluge of rain had turned the nearly vertical trail into a morass of mud and loose stones. I didn't know this, however, because I had to stop momentarily to help one of the hikers with a problem she was having. Meanwhile the group forged ahead. When I caught up to them, several of the hikers were strung out at different heights on the Indian Head Trail looking uncertain about whether they should keep going up. There was no uncertainty about it for me. I ordered everyone back down, and we returned to the Lake Road on the Gill Brook Trail, seeing all the waterfalls in reverse. We got back to the inn several hours earlier than expected, but this was one of those cases where I don't believe anyone complained. Too much water, even if it pumps up the waterfalls, is not a good thing.

One of many falls on Gill Brook.

To Gill Brook Trailhead: From the gatehouse, begin following the 3.5-mile-long Lake Road/Trail south, immediately passing under the famous handcrafted rustic gate—a 1986 replica of the original Lake Road gate. A number of trails to the right emanate from the road early on, but stay the course and don't be distracted.

In my early Keene Valley hiking days, I clearly remember paying money to ride the Ausable Club bus from the Ausable Club to Lower Ausable Lake. Club members had first dibs, but if there was room left, nonmembers could buy a ticket, the price of which changed over the years from $2.00 to $5.00 and then higher. This practice was discontinued in 1995, probably because it became burdensome to the club. The only vehicle today that is allowed on the Lake Road is the Ausable Club bus. Even bicycles are not allowed—nor dogs either, for that matter. The obvious intent is to try to prevent overuse of this section of the High Peaks, but judging by the number of hikers who sign in at the gatehouse on a busy weekend, this attempt is proving less than successful.

44. THE FLUME

At 1.1 miles from the gatehouse you will cross over a vehicular bridge spanning Gill Brook (44°08.305′N 73°47.876′W). From this point on, Gill Brook is always to your left.

In a couple of hundred feet from the bridge, an unmarked path to your left leads to The Flume (44°08.186′N 73°47.961′W)—a narrow-cut, 10–15-foot-

Gill Brook Flume. Photograph by John Haywood.

deep, ~60-foot-long chasm containing an 8-foot-high cascade at its head. Elizabeth Jaffe and Howard Jaffe, in *Geology of the Adirondack High Peaks* (1986), write, "This steep gorge was caused by the weathering-out of a narrow (1–2′, 30–61 cm) dike."

You can either follow this 0.1-mile-long spur path or stay on the Lake Road for another 0.1 mile and then bear left where a green sign indicates "Flume." The top of The Flume is virtually at roadside here.

In order to look into The Flume from the level of the stream, you will need to descend to the north end of The Flume, cross over to the other side of the stream if you can, go around a pool of water, and then scamper up to a rock outcrop from where you can peer around into the interior of the chasm. The ability to do this may be water-level dependent. You can also look down into the interior of The Flume from its west rim, but the view is decidedly less spectacular—especially if you wish to take photographs.

A picture of The Flume can be seen in Edith Pilcher's 1987 book, *Up the Lake Road*.

Back on the Lake Road, an 8-foot-high dam (44°07.942'N 73°48.321'W) to your left is passed after another 0.4 mile. Take note of a pipe extending from the dam that follows the east bank downstream. You may have noted its presence at times during the hike along the Lake Road.

At 1.8 miles from the gatehouse (or 0.6 mile from where The Flume loop reemerges), you will come to the red-blazed Gill Brook trailhead on your left, where a sign indicates "Gill Brook. Colvin. Indian Head" (44°07.843'N 73°48.413'W).

45. GILL BROOK STEPS

A series of small cascades on a rising slab of bedrock can be seen at the very beginning of the Gill Brook Trail. The Northern New York Waterfalls Web site refers to these cascades, I believe, as Gill Brook Steps (44°07.843'N 73°48.413'W). Just upstream is a 3-foot-high block-shaped cascade, which is quickly followed by another 3-foot-high block-shaped cascade, perhaps considered part of the Steps.

46. ARTIST FALLS

In over 0.2 mile you will come to a 20-foot-high block fall (44°07.659'N 73°48.563'W) that is considered to be, by popular consensus, Artist Falls. The reality, however, is that no one knows with absolute certainty which of the waterfalls on Gill Brook is actually *the* Artist Falls referred to by nineteenth-century artists.

The famous nineteenth-century surveyor, Verplanck Colvin, visited Artist Falls in 1885. From a detailed account in his *Seventh Annual Report of the Progress of the Topographic Survey of the Adirondack Region of New York*, we actually know the precise date and time that he arrived at the falls — October

Artist Falls was named for nineteenth-century artists who visited it. Photograph by John Haywood.

13 at 5:45 PM. He described the waterfall as "a beautiful little cascade in the stream ... where the water pours in a clear sheet over a sloping rock into a

crystal pool, to glide brightly away amid the giant boulders below." Colvin left a benchmark on the rock, determining it to be at an altitude of 1,637 feet.

An 1893 print of a waterfall on Gill Brook entitled *Surprise Falls* could, in fact, also be Artists Falls.

A wooden cross surrounded by a circle of stones is visible next to the fall, apparently erected to memorialize a tragic event.

47. MORE CASCADES

Forging ahead, in less than another 0.1 mile, an 8-foot-high cascade elongated over a distance of 20 feet is followed by a 20-foot-high, fairly inclined cascade (44°07.541'N 73°48.551'W).

48. SPLIT FALLS

At around the 0.5-mile mark from the start of the Gill Brook Trail, you will come to a 20-foot-high waterfall (44°07.490'N 73°48.568'W) that I am calling Split Falls because of the manner by which Gill Brook is split into two rivulets by a choke stone at the top of the waterfall. A photograph of the waterfall can be seen in James Kraus's *Adirondack Moments* (2009) and in Cliff Reiter's *Witness the Forever Wild: A Guide to Favorite Hikes around the Adirondack High Peaks* (2008).

I feel that Split Falls is a much better candidate for the mantle of "Artist Falls." To me it looks like the kind of waterfall that would have appealed to a nineteenth-century artist. In this belief I am not alone. Scott Brown in his 2010 book, *New York Waterfalls*, also calls this cascade Artists Falls and includes a photograph of it. Scott and I may be hopelessly wrong on this point, but who can demonstrate with unwavering conviction that we are?

The waterfall is reached by following a "scenic" trail sign that takes you down to the cascade and then back up to the main trail a short distance farther upstream.

Split Falls is one of the more distinctive-looking waterfalls on Gill Brook.

49. UNNAMED SCENIC CASCADES

Continuing on the main trail, you will immediately find yourself walking next to the edge of a fairly deep chasm. There are no cascades to be seen in its interior, but it is a geological feature worth noting. After another hundred feet, however, a second scenic trail sign leads you to a huge area of inclined streambed that extends for perhaps 200 feet. At the top is a 30-foot-long flume containing a 15-20-foot-high waterfall (44°07.443'N 73°48.613'W).

Fall on Gill Brook.

Farther upstream, less than 0.05 mile away, is a 10-foot-high block cascade diagonally facing the trail (44°07.357'N 73°48.630'W). It is very scenic and, best of all, can be easily seen without leaving the trail.

Continuing on, you will quickly come to a pretty, 25-foot-high cascade that drops into a short ravine (44°07.309'N 73°48.644'W).

Within another couple of hundred feet, a 2-tiered, 15-foot-high cascade is encountered. It flows into a small pothole and then immediately out into a much larger, 10-foot-long, 8-foot-wide pothole (44°07.277'N 73°48.658'W). This may be the most unusual cascade on Gill Brook.

In less than 0.05 mile farther, you will reach an elongated cascade formed by a highly inclined slab of bedrock (44°07.234'N 73°48.688'W).

Then, after another 0.2 mile, Gill Brook Trail's junction with "Lake Road via Cut-off" is reached. Just 100 feet upstream from here is a 15-foot-high cascade (44°07.061'N 73°48.785'W).

Continuing uphill from the Lake Road junction, there are no more cascades visible along the Gill Brook Trail until you reach Upper Gill Brook Falls (and even then, you will not see the waterfall until you bushwhack down to it). Along the way, the upper junction for Indian Head (44°06.570'N 73°49.030'W) is passed where a trail sign points out that you have now gone

1.1 miles along the Gill Brook Trail and are 4.2 miles from the parking area near Route 73. There is also a 12-foot-high boulder across from the junction whose 6-foot overhang provides a convenient shelter.

50. UPPER GILL BROOK FALLS

From the "Lake Road via Cut-off" junction, continue your uphill ascent on the Gill Brook Trail until you reach a GPS reading of around 44°06.262'N 73°49.239'W. This is roughly 2.4 miles from the Gill Brook trailhead. On the way up you may have noticed while glimpsing scenery through trees how the ridge you are on and the one on the opposite side of Gill Brook are slowly coming together to form a huge gulf. Upper Gill Brook Falls is near the top of this gulf.

After climbing up a series of stone steps and reaching a point where the trail levels off momentarily, you will climb down 15 feet. A couple of hundred feet from this point is where the bushwhack to your left begins (approximately 44°06.262'N 73°49.239'W). If you end up reaching the junction for Nippletop (4,620') and Mt. Colvin (4,057') at a GPS reading of 44°06.144'N 73°49.397'W, then you have climbed too high and are probably 0.2–0.3 mile up the trail from where you need to be to reach this waterfall.

Head into the woods, scampering down a moderately steep slope toward Gill Brook while negotiating a considerable amount of blowdown along the way. Listen carefully and the distant sound of Upper Gill Brook Falls will guide you in. At some point you will need to make your way over to a rocky ridgeline that extends steeply downhill. You can descend with only moderate effort to the streambed below, ending up roughly 50–100 feet downstream from the base of the waterfall at a GPS reading of 44°06.240'N 73°49.180'W. The elevation here is 3,040 feet.

Less than a hundred feet upstream from the top of Upper Gill Brook Falls is another fall where the stream drops nearly vertically for 20 feet. This section is even more difficult to access.

Scrambling back up to the Gill Brook Trail shouldn't pose any major difficulty as long as you keep going straight up the side of the gorge. If for some reason you became disoriented and are not able to find the trail, the worst that can happen is that you return to the bottom of the gulf and follow Gill Brook downstream until you get to the lower cascades.

Upper Gill Brook Falls is a big waterfall—far bigger than any of the waterfalls located downstream on Gill Brook. In fact Upper Gill Brook Falls may be almost as big as the rest of them combined. I would estimate its height at 40 feet. The waterfall, although technically two-tiered, is nearly vertical, with the falling water deflected slightly to the left by a ledge about halfway down the rock face. It is enclosed in a very deep canyon, leaving hikers with no way of continuing upstream without first climbing out of the gorge and then descending to the stream again above the fall.

Upper Gill Brook Falls is nearly as large as all the lower falls combined.

When I first came to this waterfall, I thought I had reached Fairy Ladder Falls. The elevation was about right, as was the general location. What I found puzzling, however, was how someone could have named such a vertical drop "Fairy Ladder Falls." It wasn't until later, while browsing through John Winkler's 1995 book *A Bushwhacker's View of the Adirondacks*, that I saw a photograph of a waterfall labeled Fairy Ladder Falls. The waterfall I had just seen was not the one in Winkler's book.

Undoubtedly, dozens of hikers have been to this little-known 40-foot-high waterfall over the years, but I can find no record of its existence in any of the hundreds of books I have reviewed while doing research. For lack of a better name, I am calling it Upper Gill Brook Falls because it is over a mile upstream from the lower falls on Gill Brook.

51. FAIRY LADDER FALLS

Fairy Ladder Falls is historically significant, for it was visited in 1873 by the famous Adirondack surveyor Verplanck Colvin along with his assistant Mills Blake, botanist Charles Peck, and three guides—one of them being Old Mountain Phelps, the famous Keene Valley guide (see "Phelps Falls"). They camped near the fall before ascending Mount Colvin (whose name obviously comes from Verplanck Colvin and this famous ascent). It was Colvin himself who named the cascade Fairy Ladder Falls.

Fairy Ladder Falls is also significant for being touted as the highest waterfall in the Adirondacks because of its elevation. This is a claim that I will dispute later in the book.

In *Discover the Adirondack High Peaks* (1989), Barbara McMartin describes the waterfall as "ninety-foot stairs" that are gilded by "spray and foam and rainbows." Charles Dudley Warner, essayist and novelist, also wrote about Fairy Ladder Falls: "We emerged from the defile into an open basin, formed by the curved side of the mountain, and stood silent before a waterfall coming down out of the sky in the center of the curve. … It appears to have a height of something like a hundred and fifty feet, and the water falls obliquely across the face of the cliff from left to right in short steps, which in moonlight might seem like a veritable ladder for fairies."

A photograph of Fairy Ladder Falls is included in John Winkler's 1995 book, *A Bushwhacker's View of the Adirondacks*, showing a long, descending cascade with many steps and drops.

Trail junction near Fairy Ladder Falls

To reach Fairy Ladder Falls: Continue following the Gill Brook Trail up to the trail junction (44°06.144'N 73°49.397'W) for Nippletop and Mt. Colvin. Begin following the blue-blazed trail toward Nippletop. At the right time of the year, you should be able to hear the roar of Fairy Ladder Falls off in the distance as well as catch glimpses of its top through the trees. From here you are essentially on your own, bushwhacking down into the gorge and over to the base of Fairy Ladder Falls, a trek of ~0.4 mile. At one time a quasi-path apparently led part of the way to Fairy Ladder Falls, but there is no sign of its existence now.

Fairy Ladder Falls is located at a GPS reading of 44°05.987'N 73°49.242'W, according to Google Earth, and at an elevation of 3,265 feet.

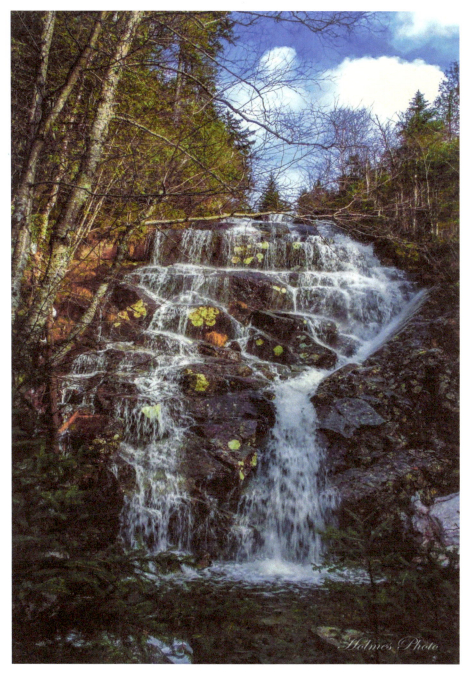

Fairy Ladder Falls is the highest waterfall on Gill Brook. Photograph by John Holmes.

V: KEENE VALLEY AREA

Hulls Falls. Postcard c. 1920.

During the latter part of the nineteenth century, Keene Valley was a Mecca for artists, painters, and writers who came to the area by the wagonloads. They called the soaring mountains and deep valley the "Yosemite of the East" and the "Switzerland of America."

Keene Valley was originally known as Keene Flats, having at one time been the flat bed of post-glacial Lake Keene, which rose up to as high as the sand terraces on East Hill northeast of Keene, about 400 feet above the present valley floor. The name permanently changed to Keene Valley in 1883.

When J.H. Mather & L.P. Brockett wrote about Essex County in their 1853 book, *A Geological History of the State of New York*, they could just as well have been writing about the High Peaks at Keene Valley: "The character of its surface is such as to produce this result; in its deep chasms and mountain gorges, its ravines and dells, bounded by walls of everduring granite, the waters which fall upon the hills, or the product of the melting snows upon its lofty peaks, gather and remain, till they have attained sufficient height to overflow the barriers which restrain them."

In *Keene Valley: In the Heart of the Mountains* (1898), Katherine Elizabeth McClellan writes, "Keene Valley has long been known as the home of the artist and haunt of the philosopher, for here nature gives with a lavish hand such a combination of wilderness and peaceful habitation as can be found nowhere else in the Adirondacks."

Because Keene Valley is uniformly flat, it has been subjected to repeated floods when the East Branch goes on its periodic rampages. During one of our annual Waterfall Weekends, we had problems accessing roads northwest of Keene Valley because of flood conditions. This was after some pretty wild weather that brought down a massive number of trees. I remember talking to a ranger at Chapel Pond who stated that, although authorities had issued a travel advisory warning hikers to stay away, he was under no illusion that the advisory would do anything but produce just the opposite effect and attract more curiosity-seekers. Many people are irresistibly drawn to witness scenes of destruction.

Today, Keene Valley is a seasonal hub of activity, attracting many thousands of people during the warmer months. But what was it like during its early days? According to the nineteenth-century writer Charles Dudley Warner, "Keene Flats on the South Branch [which we call the East Branch] of the Ausable River is a charming interval among the mountain meadows of the Ausable, a sandy road strung with farmhouses and a post office with mail twice a week. Also a school house, which serves for itinerant preachers occasionally." Very bucolic. Not much has changed in many ways, except for the hordes of hikers that now descend seasonally on the tiny community.

52. MOSSY CASCADE

Mossy Cascade (44°10.025'N 73°46.280'W) is a 40–45-foot-high waterfall formed on Mossy Cascade Brook, a medium-sized stream that rises on the south shoulder of Spread Eagle Mtn. (~2,821') and Hopkins Mtn. (~3,152') and flows into the East Branch of the Ausable River. A lumber camp was once based along Mossy Cascade Brook until operations ceased in 1924.

It's easy to see how Mossy Cascade might have gotten its name. At one time there was considerably more moss on and by the falls, but recent tropical storms have stripped away much of the green coloration.

Mossy Cascade gets harder to access with each successive tropical storm. Photograph by John Haywood.

Mossy Cascade is one of the Adirondacks' more frequently visited cascades. In fact, its image graces the cover of my first waterfall guidebook, *Adirondack Waterfall Guide*, in a photograph taken by Nathan Farb, arguably the most prominent Adirondack photographer of modern times. It was my wife Barbara who contacted Nathan Farb to ask him to allow Black Dome Press to use the picture as a cover photo. Nathan's initial response to Barbara was one of amazement, "How did you get my phone number?" Years later when I ran into Nathan Farb he stated that the book occupied a prized spot in his lavatory. This, I assume, was meant as a compliment.

Since the late 1990s a number of other highly regarded photographers have come into prominence, including Carl Heilman II, John Haywood, Mark Bowie, Nancy Battaglia, Den Linnehan, Tony Beaver, and many others, and a number of them have taken striking photos of Mossy Cascade.

Russell gets the boot: We have led hikes to Mossy Cascade on at least three occasions. On one of them, I was up ahead when I heard a cacophony of sound coming from behind me. Several of the group had gathered around one of the hikers. I rushed back as fast as I could, fearing the worst. Did someone twist an ankle or break a leg? Thank goodness, no. The sole of someone's boot had partially detached and was now flapping about like a clown's shoe. I did what any good guide would do. I reached into my backpack and pulled out the ultimate problem solver—a roll of duct tape. Within minutes her boot was wrapped up tighter than a mummy's torso and we continued on without further ado, all thanks to the magical powers of duct tape.

To get there: From Underwood (junction of Routes 73 & 9) drive northwest on Route 73 for ~6.4 miles. Park on your right just before crossing over the bridge spanning the East Branch of the Ausable River (44°09.758'N 73°46.641'W).

From Holt's Corner (junction of Routes 9N South & 73), drive south on Route 73 for ~5.1 miles. Park to your left just after crossing over the bridge spanning the East Branch of the Ausable River.

The hike up to the gorge containing Mossy Cascade is a relatively easy 0.7-mile trek. Follow the red-marked trail north, initially paralleling the East Branch of the Ausable River for the first 0.3 mile. This is an opportunity to enjoy the sights and sounds of the river below. The trail then bears right and leaves the river behind as it proceeds uphill northeast, initially following an old abandoned road called Beers Bridge Way. Mossy Cascade Brook will appear on your left.

After 0.7 mile from the trailhead, you will come to a fairly deep gorge to your left, from where Mossy Cascade Brook emerges. According to Elizabeth Jaffe and Howard Jaffe in *Geology of the Adirondack High Peaks* (1986), "In the ravine is a very pretty shattered zone of rock, trending vertically and about northeast, which is probably the cause of the falls you see here." Up until about ten years ago a fairly discernible path led into the gorge next to the streambed, ending in a scramble over the side of a 6-foot-high cascade and then up to the base of Mossy Cascade. In those early days the path was reasonably traversable, allowing Barbara and me to take a sizeable group of hikers in without any problems. More recently, however, erosion has taken its toll as one tropical storm after another has devastated the area. The path is now essentially gone.

For those venturing into the gorge today, the less than 0.05-mile-long interior hike requires edging along the east bank as best as you can and scrambling where necessary—particularly if you want to keep your boots and socks dry. As if that wasn't enough, a considerable amount of blowdown including rocks, boulders, and gravel from side wall erosion has gathered in places, creating major obstructions to get around—and all of it, of course, gets reconfigured with each new onslaught of floodwaters. I found it a bit unnerving when I hiked in this year, feeling uneasy as I looked up at a rather unstable side wall that loomed above me.

Some people are satisfied just to reach the first cascade. However, if you get this far, there is no point stopping. Mossy Cascade lies just around the bend. But suppose you don't feel like struggling through the gorge to reach Mossy Cascade? Is there another way to see the fall? Fortunately, there is, from an overlook of sorts. Continue following the main trail uphill past the gorge's entrance. The climb is fairly steep. Near the top, ignore the faint trail to your left (assuming that you even notice it at all) that heads dangerously down into the gorge. This path will lead you to nothing but trouble; descending here could prove deadly if you were to lose your footing.

When you are finally upstream above Mossy Cascade, scamper down the bank to your left. You will see a number of tiny cascades upstream from here. Cross over the streambed just upstream from the top of the fall, staying back a safe distance, and then bushwhack around the sloping rim of the gorge until you have gone several hundred feet. You will reach a point where you can look between trees into the gorge to see the waterfall below. It is not the ideal overlook, to be sure, but it has at least a partial view.

53. DEER BROOK FALLS & DEER BROOK GORGE CASCADES

On the opposite side of Route 73, south of where Mossy Cascade Brook enters the East Branch, is Deer Brook, a medium-sized stream that rises from the southwest side of Hedgehog Mtn. (3,389′) and flows into the East Branch of the Ausable River. Deer Brook contains a number of waterfalls within the Deer Brook Gorge, as well as one at a footbridge crossing and another one on

The pothole cascade and footbridge crossing on Deer Brook.

a tributary just above the footbridge. However, it is 80-foot-high Deer Brook Falls (44°09.487'N, 73°47.484'W), the uppermost waterfall on Deer Brook, that most people come to see.

Deer Brook Falls is a not only the largest waterfall on Deer Brook, but also its uppermost one.

The Deer Brook Gorge hike that I best remember was when we took the waterfallers through the gorge, over to Deer Brook Falls, and then up to Snow Mtn. (the lowest of the foothill mountains of the Great Range) where we rested and had lunch, all the while enjoying crystal clear views of Giant Mountain. The hike concluded by following the Rooster Comb/Snow Mtn.

Trail back to a manmade pond near the Rooster Comb Trailhead, and then over to Trails End Inn (our point of origin), only 0.05 mile away from the pond. This was the one time when hikers could enjoy the luxury of arriving back at their starting point without having to get into a car in order to return to home base.

A photograph of Deer Brook Falls can be seen in Scott E. Brown's *New York Waterfalls* (2010).

Death comes too early for one of our hikers: If you visit Trails End Inn, be sure to look for the memorial bench dedicated to one of our core hikers, Marjorie Ann Tierney, who died in 2003 at age 56 from a terrible accident in her New York City apartment. Right from the start, she was always there for us. Now she will always be there for us in memory.

To get there: From Underwood (junction of Routes 73 & 9) drive northwest on Route 73 for ~6.6 miles. Park on your right after crossing over the bridge spanning the Ausable River's East Branch (44°09.891'N 73°46.764'W).

From Holt's Corner (junction of Routes 9N South & 73) drive south on Route 73 for ~5.0 miles. Park to your left just before crossing over the bridge spanning the East Branch.

The hike up Deer Brook to Deer Brook Falls is totally different from the Mossy Cascade trek. You will immediately enter a gorge, following blue markers, and you will stay inside the gorge for virtually the entire hike as you head west. The trail (if there still is one, thanks to all the tropical storms) crosses the stream at least four times and takes you past one cascade after another. This is a waterfaller's paradise, for not only do you encounter a series of waterfalls, but they are all framed by the gorge's steep walls and glistening rocks, making the hike both scenic and wild.

At 0.1 mile from the trailhead, you will pass over a private driveway. It connects to a dirt road to your left that heads uphill, paralleling Deer Brook Gorge, for 0.6 mile until it is rejoined by the Deer Brook Path coming up from inside the gorge. The dirt road is referred to as the High-Water Route and is used by hikers when too much water is flowing through the gorge. It may now be the route that many people choose to reach Deer Brook Falls.

We will assume, however, that you wish to make your way through the entirety of the gorge. Your efforts will be well rewarded, for there are a number of small-to-medium-sized cascades to see along the way. In fact, they're not to be missed.

Near the end of the gorge, you will also encounter a cave to your right just before the last stream crossing. It is quite unusual for an Adirondack cave, for it was created when a huge section of rock and earth slid out from underneath the upper, rocky part of the side wall.

Once you come out of the gorge after 0.7 mile, continue west along the rim trail (an abandoned continuation of Deer Brook Way) for another 0.1 mile. When you reach the point where the main trail crosses over Deer Brook and continues on to Snow Mtn. (2,375') and Rooster Comb Mtn. (2,788'), take note of a distinctive cascade with a large pothole by the bridge—one of the more memorable sites along the hike.

Instead of crossing over the footbridge, continue straight ahead, following a spur trail usually marked "Deer Brook Falls" that parallels the south bank of Deer Brook. Almost immediately you will observe a sizeable cascade to your right formed on a tiny tributary that falls over the side wall of Deer Brook. It will be directly facing you. This is a seasonal cascade, so you may not notice it if you are hiking later in the season. Continue upstream on the spur trail for less than 300 feet to reach Deer Brook Falls—an 80-foot-high, monster cascade.

The hike through the Deer Brook gorge can be full of surprises. Each time there is a major deluge of water—particularly from tropical storms like Irene and Sandy—the gorge is significantly altered. At best the hike should be considered a bushwhack, even though sections of trail may still remain or have been partially restored.

For those not wishing to undertake the route through the gorge, all you need do is follow Deer Brook Way (a road) uphill to its end and then continue on the rim via the W.A. White Connector Trail (named after one of the founders of the Adirondack Trail Improvement Society). Unfortunately, you will bypass many of the pretty cascades, but this route will still get you to the main waterfall, the pothole cascade by the footbridge, and the seasonal cascade formed on the tiny tributary.

54. ROADSIDE FALLS

This 15-foot-high waterfall is formed on a small unnamed stream that rises from the south shoulder of Snow Mtn. (2,375') and flows into the East Branch of the Ausable River immediately after being channeled under Route 73. In all likelihood no waterfall existed here until engineers blasted out the side of a hill when constructing Route 73. What was then a cascading stream turned into a cascade. Roadside Falls (a name I am giving it) is both pretty and accessible. You can view it either while driving by on Route 73 or by turning into a large pull-off just south of the cascade and then walking up to it.

Roadside Waterfall is the most visible waterfall along Route 73, owing its existence to highway engineers who blasted away rock while widening the road.

Roadside Falls is best enjoyed in the spring or early summer when water is still flowing. The last time I drove by, only wet rock was visible. In the early spring, though, this waterfall puts on quite a show and is impossible to miss as you drive along Route 73. The fact that this unnamed stream doesn't show up on most topographical maps is evidence of its meager size.

A photograph of the waterfall can be seen in Mark Bowie's *Adirondack Waters: Spirit of the Mountains*, where it is listed as "Seasonal Waterfall, Keene Valley."

About 0.1 mile upstream, the creek drops 35–40 feet down through rocks and boulders, producing a medium-sized, ill-defined cascade. Unfortunately, this upstream cascade is located on posted land, so remain on Route 73 and enjoy the roadside show.

To get there: From Underwood (junction of Routes 73 & 9) drive northwest on Route 73 for ~7.1 miles (or 0.5 mile from the trailheads for Mossy Cascade and Deer Brook Falls). Look for the cascade on your left. The pull-off is directly to your right (44°10.217′N 73°46.944′W).

From Holt's Corner (junction of Route 73 & 9N South), drive south on Route 73 for ~4.5 miles. The cascade will be on your right.

55. FLUME FALLS

Flume Falls consist of a series of 6–8-foot-high cascades formed on Flume Brook, a small stream that rises from a col between Hedgehog Mtn. (3,389′) and Rooster Comb Mtn. (2,788′) and flows into the East Branch of the Ausable River.

These falls are contained in a narrow ravine with lopsided walls—the north side wall being considerably higher and more straight-cut than the south side wall (approximately 44°10.713′N 73°47.289′W). A quick look at a topographical map reveals a unique feature about Flume Brook—it races down the mountainside as straight as a ruler's edge, cutting diagonally across the landscape.

Flume Brook has also been known as Washbond's Flume Brook, after Henry Washbond, who ran the Flume Cottage near the flume. In addition it has been called Sachs Flume Brook after Dr. Ernest Sachs, a neurosurgeon who built his house on the former site of Flume Cottage. There are other names as well: Snow Mountain Brook, based on the fact that the stream rises from Snow Mountain; and Rushing Brook, the name of the short road (Rushing Brook Way) that parallels the stream.

Until 1998 the area by the flume was accessible to hikers, who were able to enjoy this natural phenomenon and its cascades. Circumstances, however, forced the Adirondack Trail Improvement Society (ATIS) to reroute the trail to its present location farther north, diverting traffic away from a large summer home near the top of the north wall of the gorge just uphill from where a flight of stone steps leads down to the streambed below the flume. This is a private area, so keep your distance.

It is possible to access the upper part of Flume Brook if you wish. A section of the trail to Snow Mountain parallels the brook for roughly 0.5 mile. It can be reached by hiking up the Rooster Comb/Snow Mountain Trail from the valley for ~0.7 mile. When the trail bears right at Flume Brook, you have arrived at the point where the old Sachs's Flume Brook Trail used to come up. I have explored this upper portion, which involves a bushwhack paralleling the Rooster Comb/Snow Mtn. Trail, but there are no major cascades to be found—only minor ones.

56. HOPKINS BROOK FALLS

Hopkins Brook Falls consist of several small-to-medium-sized cascades on three different sections of the stream.

Hopkins Brook and Hopkins Mountain (3,183′) were named for Reverend Erastus Hopkins, a Troy minister who hiked extensively in the area. The Ranney Way Bridge, which spans the East Branch and must be crossed to reach the falls, is a 1902 iron Pratt Pony Truss bridge that was originally located in New Russia. It was moved to its present location around 1925. Ranney Way Road and Ranney Way Bridge are attributed to one of the Ranney family who lived in Keene Valley.

Although it's unlikely that the falls on Hopkins Brook have ever been industrialized, that doesn't necessarily mean that their waters haven't at some point been used. Look closely as you follow the trail uphill and you will see remnants of pipe both on top of the ground and going into the ground. It seems reasonable to assume that this conduit once brought water down to the camps below.

Getting there: From Underwood (junction of Routes 73 & 9) drive northeast on Route 73 for ~8.2 miles and turn left into the large parking area for Rooster Comb and Snow Mountain.

From Adirondack Street in the center of Keene, drive southeast on Route 73 for 0.4 mile and turn right into the Rooster Comb/Snow Mountain parking area (44°11.125′N 73°47.198′W).

From the parking area, walk across Route 73 and continue south for several hundred feet until you come to Ranney Way Road. Turn left, cross over the bridge, and follow the road southeast for 0.2 mile. When you come to a fork, bear left and proceed northeast for another 0.1 mile. Along the way a number of camps will be passed. You will soon come to the end of the road, where the blue-blazed trail to Hopkins and Giant Mountain begins to your right (44°10.974′N 73°46.861′W). The trail is clearly marked and follows an old abandoned road uphill. Head east. Soon you will begin paralleling the stream, to your right. In less than 0.2 mile you will come to a tiny, 4-foot-high, block-shaped fall.

Several pretty waterfalls have formed on Hopkins Brook.

At 0.3 mile you will see a series of pretty drops and cascades totaling 30 feet in height. They are visible from the trail. From here, bushwhack downstream for a couple of hundred feet to see a pretty, 10-foot-high plunge fall. Take note of a 2-inch-diameter pipe that parallels the stream along the top of the bank.

Continue east on the trail, heading farther uphill. At 0.4 mile you will see a Forest Preserve sign. Immediately after the sign, the trail crosses Hopkins Brook, pulls away momentarily from the stream, and then returns as the uphill climb steadily increases. After 0.1 mile from the stream crossing, you will see several large boulders to the left of the trail. Downstream, below, is an elongated 20-foot cascade that can only be appreciated if you scramble down to the bottom of the ravine.

Continuing on the blue-blazed trail for another 0.2 mile, you will see off to your left a 15-foot-high, fairly nondescript cascade. It lies 30 feet from the trail, formed at the confluence of two tiny streams.

From here the trail continues upward, but there are no more waterfalls to be seen.

Beaver Deceiver

Beavers are nature's tireless engineers, makers of waterfalls. They use tree limbs, debris, and mud to dam up streams and rivers to create impoundments with water cascading over them. It seems only fitting, then, that we give the beaver honorable mention in a book about waterfalls.

So, while you're at the parking area for the Rooster Comb/Snow Mtn. trailhead, take a look at the upstream (west) end of the culvert under Route 73. You will see a trapezoidal-shaped crate in front of the drainpipe. Highway engineers, it seems, weren't able to prevent beavers from erecting dams in front of culverts until they hit upon the idea of using a trapezoidal-shaped barrier. Apparently beavers cannot figure out how to deal with a shape that doesn't exist in nature. It was after this discovery that someone came up with the clever name Beaver Deceiver.

57. PHELPS FALLS

[handwritten: NO PRIVATE]

Phelps Falls is a 30–40-foot-high cascade formed on Phelps Brook, a small stream that rises from the northeast shoulder of Hopkins Mtn. and flows into the East Branch of the Ausable River at Keene Valley. Topographical software places the fall at around 44°12.082'N 73°46.471'W. The waterfall is named for Orson Schofield Phelps (1817–1905), better known in his later years as Old Mountain Phelps—the most famous Adirondack guide to emerge from the nineteenth century. Phelps's fame was due in large part to writers like E.R. Wallace and Charles Dudley Warner, who transformed Phelps into almost a mythic woodsman. Phelps was featured in a piece written by Warner for the 1878 *Atlantic Monthly* entitled "The Primitive Man." Whether Phelps took this title as a compliment is unknown. In his 1874 book *Adirondacks Illustrated*, Seneca Ray Stoddard wrote about an encounter with Phelps near Phelps Falls: "We found him at his home near the falls that bear his name—a little old man, about five foot six in height—muffled up in an immense crop of long hair, and a beard that seemed to boil up out of his collar band; grizzley [sic] as the granite ledges he climbs, shaggy as the rough-barked cedar."

Phelps was a resident of Keene Valley and lived for a major part of his life on Beede Road, directly off Route 73 near the edge of the valley. Next to Beede Road is Phelps Brook, earlier known as Clough Brook after Parley & Esther Clough, who owned a house nearby and operated a small sawmill roughly 300 feet up from where Beede Road crosses Phelps Brook. In 1860 Phelps purchased most of the property along the north side of Beede Road from Clough as well as the sawmill, which Phelps took over and operated.

Phelps Falls is named after Orson Schofield Phelps, the famous Adirondack guide who lived nearby.

Phelps visited the waterfall frequently. It was only a stone's throw from his house. A photo used by Bill Healy on the cover of his book *The High Peaks of Essex: The Adirondack Mountains of Orson Schofield Phelps* shows Phelps standing in front of this waterfall. What a rare picture indeed it is that contains both the man and the waterfall named for him.

Unfortunately, Phelps Falls is located on private property today and cannot be accessed without special permission, nor can it be seen from roadside no matter how you crane your neck for a look. Waterfallers will just have to take pleasure in knowing that Keene Valley's most famous resident has a waterfall named after him.

By good fortune we finally did get to visit Phelps Falls after getting permission from an exceedingly gracious woman named Elizabeth, whose camp was directly downstream from the waterfall. Imagine our surprise and delight when we were invited into Elizabeth's camp and saw on her fireplace mantle a photo of Orson Phelps at Phelps Falls.

Phelps's name has not been forgotten, nor is it likely to be. The name is firmly embedded in the nomenclature of the Adirondacks. There is not only Phelps Brook and Phelps Falls, but also Phelps Mountain (4,161') and a second stream called Phelps Brook that runs off the southern flank of Phelps Mountain, undoubtedly with its own set of cascades. Even Phelps himself has not gone away in body or spirit. He lies buried in the Estes Cemetery off Beede Road (a photo of the tombstone can be seen in Tim Rowland's *High Peaks: A History of Hiking the Adirondacks from Noah to Neoprene*) and he is immortalized in a painting by Winslow Homer entitled *Two Guides* (which was used on the cover of *Adirondack Trails with Tales*, a book I coauthored with my wife, Barbara Delaney, in 2009).

58. TOWN RIDGE LOOP TRAIL FALLS

A 15-foot-high, fairly inclined, seasonal cascade has formed on an unnamed stream that rises from the lower shoulder of Blueberry Mountain. After tumbling through a jumble of boulders, the stream flows past the 1.3-mile-long Town Ridge Loop Trail and eventually into the East Branch of the Ausable River.

The upper part of the Town Ridge Loop Trail, which is shaped like a lollipop, overlooks the Marcy Air Field—a huge expanse of flat land laid down by a post-glacial lake. The land was turned into a tiny airstrip in 1940 by Alphonzo Goff, a medical doctor who was also a pilot.

To get there: From Underwood (junction of Routes 73 & 9) drive northeast on Route 73 for ~10.5 miles and turn left onto Airport Road. Take note that earlier you will pass by the south end of Airport Road.

From Holt's Corner (junction of Routes 73 & 9N South) drive south on Route 73 for ~1.0 mile and turn right onto Airport Road.

From either direction proceed west for less than 0.1 mile and turn right into the parking area for the hiking trails to Blueberry Mtn. (2,920'), Porter Mtn. (4,059'), and the Town Ridge Loop Trail (44°13.112'N 73°47.406'W).

From the west end of the parking area, walk uphill following an abandoned road. You will pass by the trailhead for Porter Mountain /

Blueberry Mountain in less than 0.1 mile. In another 0.05 mile you will pass by a 3–4-foot-high cascade on your right just downstream from a small breached dam. Look to your left for an abandoned, box-shaped cement shed with its door lying on the ground—an area obviously abundant with its own history.

Town Ridge Loop Trail Falls.

    A tiny footbridge takes you across an unnamed creek. Once on the other side, bushwhack upstream for a hundred feet to reach the base of the cascade (44°13.205′N 73°47.653′W). The waterfall can also be seen through trees along the trail.
    If you wish to complete the Town Ridge Loop Trail, return to the trail and continue uphill for another 0.1 mile to reach the 0.5-mile-long lollipop loop. There are views of the valley and airfield from an outcrop at the apex of the loop where a bench faces Route 73.

59. BLUEBERRY FALLS

This fairly nondescript cascade consists of 20 feet of falling waters over a section of exposed bedrock. The unnamed stream containing the cascade rises from the south shoulder of Blueberry Mountain (2,910') and flows into the East Branch of the Ausable River near the southeast corner of Marcy Airport. The waterfall is named for its location on Blueberry Mountain.

What I remember most about this particular hike is that one of our waterfallers tumbled into the stream while bettering his position to photograph the cascade. He didn't get hurt, but he certainly got wet. Even the little falls can get you if you're not careful.

Blueberry Falls.

To get there: See directions to the Town Ridge Loop trailhead (44°13.112'N 73°47.406'W). From the parking area, head west up an abandoned road for less than 0.1 mile and turn left onto the Blueberry Mountain / Porter Mountain Trail. The path was cut in the 1930s by Cecil Parker. The first 0.1 mile of the trail is steadily uphill. By 0.2 mile the path begins to level off. Cross over a small stream at 0.4 mile; this is the creek that contains the cascade. A huge outcrop of rock is on your left as you continue to follow the trail, which now heads steeply uphill, paralleling the stream to your right. In a hundred feet or so make your way over to the creek to reach the base of the cascade (approximately 44°12.924'N 73°47.903'W). Keep your expectations low. This is not one of the big ones, although the waterfall does look very pretty in the early spring.

60. HULLS FALLS

Having already seen a series of waterfalls on the upper section of the Ausable's East Branch in the Adirondack Mountain Reserve, we now travel to Hulls Falls, a 20-foot-high, hulking waterfall formed on the midsection of the East Branch. It was named after Eli Hull, an early settler from Killington, Connecticut. Here the river flows through a post-glacial channel that is severely cut. An earlier, pre-glacial channel that carried the river eastward around 1,443-foot-high Beede Hill became occluded by a delta of sand, forcing the East Branch onto its present course. According to Elizabeth Jaffe and Howard Jaffe in *Geology of the Adirondack High Peaks* (1986), it was the "erosion of these dikes, and the concomitant jointing, [that] probably led to the formation of the waterfall here."

Around 1820, Joseph and Alden Hull established a forge just downstream from the waterfall. Other mills would soon follow in the years to come. Little of any significance has survived from these early years except for the Hulls Fall House, a historic 1890 Queen Anne–style structure that was one of the first houses to be built in Keene.

In Dennis Squire's *New York Exposed: The Whitewater State Volume I* (2002), a photograph taken by Mike Duggan shows a kayaker going over the fall. A postcard photo of the fall can be seen in Scherelene L. Schatz's 2008 book *The Adirondacks: Postcard History Series*.

To get there: From Underwood (junction of Routes 73 & 9) drive northeast on Route 73 for ~11.1 miles (or 2.6 miles north from the junction of Route 73 & Adirondack Street in Keene Valley) and turn left onto Hulls Falls Road.

From Holt's Corner (junction of Routes 73 & 9N South) drive south on Route 73 for 0.5 mile and turn right onto Hulls Falls Road.

From either direction, head northwest on Hulls Falls Road for 0.8 mile and park next to the bridge spanning the East Branch (44°14.147′N 73°47.735′W). Hulls Falls is directly below the west side of the bridge. At one time it was possible to scamper down to the base of the fall, but those days are gone. The posted signs are not there to be disobeyed.

Hulls Falls, like Split Rock Falls near New Russia, has been a magnet attracting swimmers and sun-worshippers, and there have been a number of accidents at the fall.

Before leaving Hulls Falls behind, a brief story is worth telling. About a year or two after Black Dome Press published *Adirondack Waterfall Guide* in 2003, I was contacted by a very sweet woman who was absolutely convinced that Hulls Falls was gone because she had followed my directions and had failed to locate the waterfall. Perhaps, she suggested, the waterfall had been destroyed by a natural disaster like an earthquake, causing the waterfall to collapse much like a building does when shaken too hard.

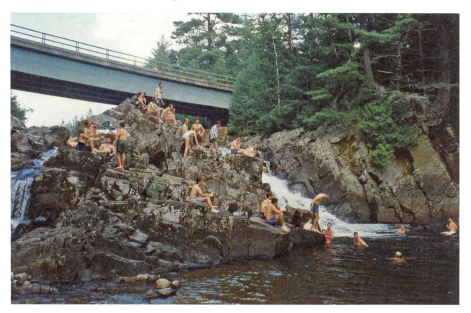

Swimming is no longer permitted at Hulls Falls.

The reality is that waterfalls don't just disappear like that. Even if the front of the waterfall collapsed, there would still be the same height differential between the streambed above and below. It just couldn't happen the way she imagined it. But, naturally, I was still relieved to see that the waterfall was there when I next visited it.

Waterfalls, to be sure, are not immune to destructive forces. In fact, they initiate them. Every second, every waterfall on this planet is breaking apart the terrain uphill from it, dismantling mountains and hills one pebble, stone, or boulder at a time. Give waterfalls several thousand years, and some of the big ones will vanish as they self-destruct, turning the streambed into one long, inclined gully. Fortunately, Hulls Falls has a long way to go before it carves itself out of existence.

For those interested in a nature walk along a 0.2-mile-long trail that leads upstream from Hulls Falls, walk over to the southeast side of the bridge where a sign states "Private Property. No fires. Camping. Littering. Walkers Welcome." It is a pleasant trek with pretty views of the river, but don't expect to see any cascades.

Returning to Route 73, farther north: From Hulls Falls, continue driving north along Hulls Falls Road, paralleling a fairly substantial gorge that the East Branch has carved out. Keep the odometer set at 0.8 mile when you start off.

At 1.5 miles you will come to a bridge connecting Hulls Falls Road with Gristmill Road. Because recent tropical storms have weakened the bridge, it is used strictly for pedestrian traffic now. Great views of the rocky streambed can be obtained here.

At 2.0 miles, Hulls Fall Road narrows momentarily and becomes a one-lane highway for the next fifty feet. You will see to your left how erosion has stripped away a sizeable portion of the embankment, threatening to claim the road itself. You will find great views into the gorge from here, but you can't stop and linger for too long.

At over 2.4 miles you will return to Route 73 in the center of Keene.

61. HULL BASIN BROOK FALLS

Barbara and I have not actually been to this waterfall, although we know of its existence and can identify its location pretty accurately on a topographical map. We even have a large photographic print of the waterfall, taken by Nathan Farb, that is hanging in our summer camp. The problem with Hull Basin Brook Falls is that you may need to cross private land to get to it, but we have not been able to verify if this is true or whether there is a legal way to access the waterfall.

Hull Basin Brook Falls is formed in an extremely narrow flume on Hull Basin Brook, a small stream that rises on the west shoulder of Blueberry Mtn. (2,920′) and flows into the East Branch of the Ausable River just below Hulls Falls. The flume lies roughly 0.5 mile upstream from the creek's confluence with the East Branch.

A photograph of the fall can be seen in Nathan Farb's 1986 book, *The Adirondacks*, as well as in his 1989 book, *100 Views of the Adirondacks*. Farb is not the only one to showcase the flume. In his 1988 book, *The Adirondacks Forever Wild*, George Wuerthner includes a photograph of the fall, which he calls Hull Brook Flume. Wuerthner writes, "Flumes are created when the basement rock is intruded by a softer rock. Because it is softer than adjacent material, it is more rapidly eroded by water and hence creates narrow gorges."

62. CHAMPAGNE FALLS & GRISTMILL ROAD GORGE

No PRIVATE

Champagne Falls (44°15.056'N 73°47.799'W) is a 6-foot-high drop on the East Branch of the Ausable River. It is contained in an impressive gorge where swift-moving waters, small cascades, and rapids have created a dynamic vista.

Take note that the property overlooking the falls is privately owned and posted. A sign drives home the point of just how dangerous swimming in the gorge can be. On July 22, 2010, tragedy befell a 12-year old boy from Florida while swimming with his dad by the falls. Although the Keene Volunteer Rescue Squad responded quickly and had the boy out of the water within 15 minutes, he died at the Fletcher Allen Health Center in Burlington. So stay safe and just look.

A photograph of an unidentified waterfall on the East Branch in Keene, probably Champagne Falls, can be seen in Bradford Van Diver's *Field Guide to Upstate N.Y.* (1980).

Just upstream from Champagne Falls is a rocky, 0.1-mile-long section of the gorge (44°15.020'N 73°47.879'W).

There has been a fair amount of past industrial activity here both in the gorge and the flume area. Listed in order going upstream, they were: David Graves's forge (c. 1820); Sylvanus Wells's sawmill (c. 1820); Israel Kent's gristmill (c. 1830); and Nathaniel Sherburne's gristmill (c. 1830). All total, six mills operated between Hulls Falls and the hamlet of Keene.

Over 1.0 mile upriver, the 1890 Walton Bridge connects Gristmill Road with Hulls Falls Road. In 1990 the bridge was closed because of unsafe conditions, but it has remained opened to pedestrian traffic. In 2011 it was nearly destroyed by Tropical Storm Irene, so how long it will continue to survive is anyone's guess.

Champagne Falls needs to be viewed with caution. People have died here.

To get there: From Keene (junction of Routes 73 & 9N North) cross over the bridge spanning the East Branch of the Ausable River and turn immediately left onto Gristmill Road. Head southwest for 0.4 mile to see Champagne Falls and several smaller falls that are visible from roadside, but just barely (44°15.074′N 73°47.820′W). Posted signs are very prominent at a pull-off on your left that has been blocked by large boulders. One sign states, "No swimming. Dangerous rocks and currents. A boy drowned here." Do not park or swim here. The area is regularly patrolled by the State Police, and you are only inviting trouble if you do more than take a look from the road.

VI: JOHNS BROOK

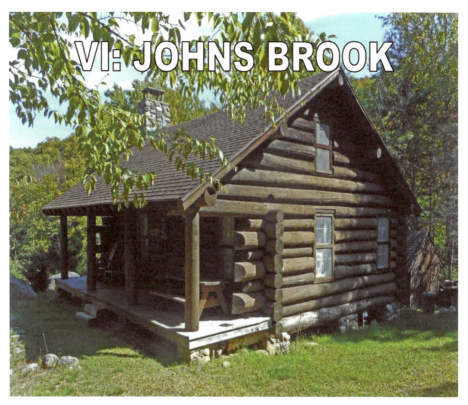

The Interior Outpost Ranger Station, one of four in the High Peaks Region.

Johns Brook is a substantial stream that rises from a col between Little Marcy Mtn. (~4,727') and Little Haystack Mtn. (~4,674'). It has an average gradient of ~250 feet per mile. The brook was named after John Gibbs, who lived at (or at least owned) the property at the confluence of Johns Brook and the East Branch of the Ausable River in 1795. The site is now occupied by The Mountaineer, a purveyor of fine outdoor equipment since 1975.

A sawmill, possibly owned by David Heald, once operated on Johns Brook ~0.2-mile upstream from its confluence with the East Branch. In 1869 a more substantial sawmill was built, probably close to where the Johns Brook Lane crosses Johns Brook.

The ravages of Tropical Storm Irene on Johns Brook were most visible down by Route 73. The Johns Brook Bridge on Route 73 was clogged by debris and its abutments were threatened. Water was everywhere. The Department of Transportation responded by carving out a new channel to prevent the bridge from being undermined, and bulldozed nearly a mile upstream, piling boulders high above the bank on both sides of the stream.

Not all were happy with this action. According to U.S. Fish & Wildlife specialist Carl Schwartz, engineers may have inadvertently created a "water cannon," where the trapped waters, under force, could fling huge boulders at the bridge. That's certainly something to picture in your mind as you follow Johns Brook upstream driving up to The Garden, one of the main entry points into the High Peaks Region.

The Phelps Trail, heading south from The Garden, was laid out in 1871 by Ed Phelps, son of Old Mountain Phelps (Orson Schofield Phelps), and Seth Dibble. Initially the route was not as popular as it is today. Nineteenth-century hikers often chose to approach Mount Marcy (5,344) and other high peaks from Lower Ausable Lake, where the views along the trail were considered more spectacular. This later changed as the forests grew back and the views became more obscured.

The Phelps Trail has been known by other names as well: the John Brook Trail after Johns Brook (which it parallels); and the Northside Trail (in recognition that it is on the opposite side of the river from the Southside Trail).

I have hiked along this trail a fair number of times, not only to waterfalls but to mountains as well. It is an extremely well-worn, heavily used path. Unfortunately it never gets even remotely close to Johns Brook until you reach the junction of the Interior Outpost Ranger Station and Johns Brook Lodge trails. At best you may occasionally hear the sound of Johns Brook in the distance.

Interestingly, whitewater enthusiasts have paddled down Johns Brook during times of high water. In his 2002 book, *New York Exposed: The Whitewater State. Volume 1*, Dennis Squires writes, "The river is big enough to start way up at Bushnell Falls," but then adds, "—but nobody has been ambitious enough to start that high yet." Paddlers, however, have put in as far upriver as Johns Brook Lodge.

Undoubtedly after hiking along the Southside Trail and seeing all the boulders, rocks, ledges, and waterfalls, you may be left wondering how paddlers do it. It's easier to understand if you are a whitewater paddler yourself.

To get to the trailhead: From Underwood (junction of Routes 73 & 9) drive northeast on Route 73 for 8.6 miles. Once you arrive at the center of the hamlet of Keene Valley, look for a DEC sign that reads "Trail to the High Peaks" and turn west onto Adirondack Street (which soon turns into Johns Brook Lane). In 0.6 mile you will cross over Johns Brook, where major rapids and boulders abound in the streambed. Be sure to look for a large split rock on your right just before crossing the bridge that spans Johns Brook.

The Garden—a good-sized parking area—is reached at the end of the road after 1.6 miles (44°11.344'N 73°48.956'W). A modest parking fee is charged. The Garden was built in 1970 and acquired its name from a vegetable garden that the parking lot replaced. Despite having room for up to 60 cars, The Garden tends to fill up quickly during busy weekends. Should you drive up to it and find that there is no room for parking, you will have to return to the village and either come back on foot (adding on another 1.6 miles each way to your hike) or catch a shuttle if one is operating on the day of your visit. No parking is allowed along the road leading up to The Garden.

PHELPS TRAIL

63. DEER BROOK CASCADES

Two footbridges cross over a stream that has been split into two channels at this point. The rivulet on the east channel consists of a 30-foot-high cascading stream, where most of the water from Deer Brook is diverted. A couple of tiny, boulder-produced cascades can be seen here. The rivulet on the west channel is slab-like but fractured into sections. Little water is carried through here, but under conditions of massive water flow, it's possible that a slab fall may form.

The main falls are 200 feet upstream, not far from the Deer Brook lean-to. A 12-foot-high cascade is immediately followed downstream by a 25-foot-high, 30-foot-long waterslide (44°10.690'N 73°50.028'W).

It should come as no surprise that two Deer Brooks can be found within a dozen miles of each other and that both are waterfall-bearing (see the write-up of the other "Deer Brook Falls" near St. Huberts). Deer are very common in the woods, and so are streams that bear their name.

To Deer Brook Cascades: From The Garden, follow the yellow-blazed Phelps Trail southwest for 1.3 miles (44°10.643'N 73°50.032'W). You will come to Deer Brook, where two single-railed footbridges, 50 feet apart, cross the brook below where the stream splits into two rivulets. Just downstream from the footbridges, the two rivulets rejoin and become one stream again. You can view the streams coming back together from atop a mound 10 feet downhill from the footbridges.

Deer Brook Falls is the first cascade encountered along the Phelps Trail.

Along the hike, take note of a number of massive glacial boulders dotting the landscape, several of which are trailside. One of them is called Resting Rock, which Neal Burdick, in *Adirondack Life's 2007 Annual Guide to the Great Outdoors*, describes as about halfway along the trail to Johns Brook Lodge. It is a flat-surfaced boulder with a U.S. Geological Survey benchmark embedded in it (44°10.242'N 73°50.465'W).

64. UPPER JOHNS BROOK FLUME & FALLS

Near the Interior Outpost Ranger Station several falls have formed in a 100-foot-long flume lined with 20-foot-high walls. The chasm is spanned by a suspension footbridge near its head that was recently installed by the NYSDEC, replacing an older one.

The upper flume on Johns Brook is spanned by a suspension bridge, creating dynamic views of the gorge.

Immediately downstream from the suspension bridge is a 6-foot-high cascade followed by a 6–8-foot-high cascade. A hundred feet or so upstream from the bridge is a 6–8-foot-high fall entering diagonally from the left where an island has split the stream into two channels momentarily. Only part of the cascade can be seen from the bridge.

A photograph of the falls can be seen in Poncho Doll's *Day Trips with a Splash: Northeastern Swimming Holes* (2002).

The Johns Brook Interior Outpost is one of four Interior Outposts in the Adirondacks. All are located in the High Peaks Region and were built by the DEC's Land & Forest Division. The Johns Brook Interior Outpost Ranger Station was constructed in 1948. Improvements were subsequently made, ending with a complete overhaul in the 1960s. The other three outposts are located at Raquette Falls, Lake Colden, and Marcy Dam.

To Upper Johns Brook Flume & Falls: From The Garden, follow the yellow-blazed Phelps Trail southwest for ~3.1 miles (or ~1.8 miles from Deer Brook Cascades), paralleling Johns Brook.

When you come to the junction for the DEC Interior Outpost Ranger Station and Johns Brook Lodge, turn left, following a red-blazed trail that leads quickly to the DEC Interior Outpost Ranger Station. From here continue on the trail for another couple of hundred feet to reach a suspension bridge that crosses Johns Brook near the head of an impressive flume (44°09.656'N 73°51.271'W). There are views of cascades both upstream and downstream, but the downstream view is definitely the more dynamic one.

For the best view of the flume and its two lower falls, bushwhack downstream along either bank for less than 100 feet to reach areas of exposed bedrock overlooking the river. From a number of vantage points, one very close to the level of the river, you can gaze back upstream into the gorge and its falls. The views are wild and scenic, and far superior to those from the footbridge.

To obtain an unobstructed view of the upstream cascade, follow a faint path along the bank of the river from the northwest end of the bridge toward the back of the Interior Outpost Ranger Station. Within 100 feet you will come to an opening in the brush from where the cascade and river below can be clearly seen.

Additional cascades nearby on Johns Brook: Return to the Interior Outpost Ranger Station and Johns Brook Lodge trail junction. This time, instead of turning left toward the Ranger Station, bear right. Walk 100 feet south along the Phelps Trail and then turn left. In 50 feet you will come to Johns Brook, where several 2–4-foot-high cascades can be seen along an area of much exposed bedrock containing shallow potholes (44°09.616'N 73°51.361'W).

65. SLIDE MOUNTAIN BROOK FALLS

A number of cascades and large waterslides have formed on Slide Mountain Brook, a medium-sized stream that rises from a col between Big Slide Mtn. (4,240') and Yard Mtn. (4,009') and flows into Johns Brook just east of the trailhead. In *Discover the Adirondack High Peaks* (1989), Barbara McMartin mentions that toward the end of the hike, "The stream opens up to huge rock shelves. You walk up the slides in a small gorge. A stream with nice cascades enters from the west."

A photograph of one of the waterslides can be seen in Cliff Reiter's *Witness the Forever Wild: A Guide to Favorite Hikes around the Adirondack High Peaks* (2008).

<u>To get there</u>: From the junction of the yellow-blazed Phelps Trail and the trail leading to the Interior Outpost Ranger Station, continue west on the Phelps Trail for 0.1 mile. When you come to the sign for Big Slide Mtn. on the other side of Slide Mountain Brook, turn right and follow the trail as it leads north, heading uphill toward Big Slide Mtn. (4,240'). The waterfalls are formed along a 0.8-mile section of the trail, beginning less than 0.2 mile into the hike. From here the trail repeatedly crosses over Slide Mountain Brook.

As it turns out, when I undertook the hike this spring, the stream was running so fast and high that I was not able to rock hop across it to get to the trailhead. Even worse, the upstream high-water bridge was gone, leaving no way to cross Slide Mountain Brook safely and easily. I thought about fording the stream regardless of the conditions, but then realized I would have to do it many times over as the trail crisscrossed the brook higher up. I also realized that I was alone, without backup, the water was near freezing, and if I got swept off my feet by the current and immersed in the frigid water I could be in serious trouble.

As a result I decided to abort the hike. However, wanting to make the best of the situation by not giving up entirely, I followed a short path along the north bank that led up to the former high-water crossing. I could see the abutments where the footbridge once spanned Slide Mountain Brook. From here I did an easy bushwhack upstream along the north bank for another 0.1 mile and ended up across from the foot of an enormous waterslide coming down the mountainside from hundreds of feet above (44°09.729'N 73°51.516'W). The view was spectacular. Sometimes things work out even when they don't seem to be working out.

JOHNS BROOK LODGE

There are no waterfalls by Johns Brook Lodge (JBL), but it is a wonderful place to stay if you wish to do a stretch of sustained hiking in the High Peaks Region. Johns Brook Lodge is on one of 17 parcels of private land contained in the High Peaks Wilderness Area. The Adirondack Mountain Club purchased the land from the J. & J. Rogers Company in 1924 and opened up the lodge in 1925. Two additional camps were purchased in 1929—Camp Thistle-Do (known as "Winter Camp") and Grace Camp (which was used for lodging by volunteers). Both have since been replaced by more modern structures.

Johns Brook Lodge can accommodate up to twenty-eight guests in two, four-bunk family rooms and two large bunk rooms. It is located in the second-most-heavily-used area in the High Peaks Region. When it first opened in 1925, many of the surrounding peaks were still visible because of all the lumbering that had occurred. Those temporary views, however, ended in the late 1960s as the forest returned.

Johns Brook Lodge is 3.5 miles from the nearest road—a real wilderness retreat for backpackers.

More information on Johns Brook Lodge and the Adirondack Mountain Club (ADK) in general can be obtained at adk.org.

To Johns Brook Lodge: From the junction of the Phelps Trail and the Interior Outpost Ranger Station trail, continue south on the Phelps Trail for another 0.4 mile (or 3.5 miles from The Garden, with a total elevation change of 800 feet) until you come to Johns Brook Lodge (44°09.495'N 73°51.758'W).

66. MINOR CASCADES

No

There are a number of pretty spots along Johns Brook that contain quasi-waterfalls, i.e., small cascades produced by boulders. This one is exceptionally pretty and, best of all, lies almost next to the trail.

To get there: From Johns Brook Lodge, continue south on the Phelps Trail, following a section of trail that parallels Johns Brook and lies close to it. At ~0.7 mile from Johns Brook Lodge, the trail leads out onto the streambed and returns to the woods several hundred feet later. Soon after you will see to your left a quiet spot in the stream filled with large boulders and one or two minor cascades (44°09.131'N 73°52.335'W).

67. BUSHNELL FALLS

No

Bushnell Falls is an impressive 25–30-foot-high cascade formed on Johns Brook. The waterfall drops over an irregular-shaped rock wall into a pool of water at its base. From there the stream flows downhill another 30 feet and then down a long waterslide, dropping another 5 feet. Bushnell Falls is by far the largest waterfall on Johns Brook and also the one farthest upstream.

To the right of the fall is a large boulder and, behind it, a tremendous pile of outwash and debris that has been deposited during times of high water.

A striking photo of Bushnell Falls can be seen in Nathan Farb's 1985 classic *The Adirondacks*, as well as in his 1989 book, *100 Views of the Adirondacks*.

Bushnell Falls was named for Horace Bushnell, a minister who lived in Hartford, Connecticut, and who did some hiking in the Adirondacks. His musings were published in a number of religiously oriented books. Bushnell believed that theology's appeal was based on how it resonated with humanity's feelings, intuition, and spiritual nature. This was opposed to the prevailing orthodoxy of the day, which contended that theology's appeal was intellectual and arrived at through logical deduction. In this respect Bushnell was truly a contemporary of Ralph Waldo Emerson.

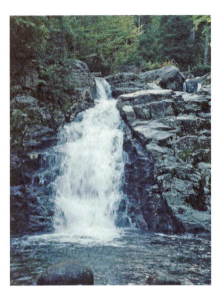

Bushnell Falls is the uppermost and largest of the Johns Brook waterfalls.

Bushnell was also a man who wasn't afraid to admit to the little annoyances in life, such as mosquitoes: "I confess there is one thing which I cannot account for—that man, who is made in God's image, can be made perfectly miserable by a little creature (indicating on the tip of his forefinger) no bigger than that."

At one time the fall and surrounding lands belonged to the Adirondack Mountain Reserve (AMR), but were gradually bought by the State of New York through a series of land purchases between 1921 and 1932.

In 1960 a group of Girl Scouts became stranded on the east side of Johns Brook near Chicken Coop Brook (formerly called Saddleback Brook), just upstream from Bushnell Falls, during a bout of unexpected high waters. After spending the night in the woods, they were rescued the next day by rangers who rigged a rope over the stream to get them across. Undoubtedly, other people have not been so lucky at Bushnell Falls, where injuries can occur. Just climbing down to it could be hazardous.

I have read that a waterfall exists on Chicken Coop Brook, but I cannot confirm its existence, having been unaware of its presence the last time I visited Bushnell Falls.

To Bushnell Falls: From Johns Brook Lodge, follow the yellow-blazed Phelps Trail south for another 1.8 miles (or a total of 5.3 miles from The Garden). When you come to the junction for "Mt. Marcy, Van Hoevenberg Junction, and Adirondack Loj," turn left and follow a 0.1-mile-long, heavily eroded, root-and-rock-strewn, blue-marked trail downhill. It comes out onto the bedrock just below Bushnell Falls (44°08.735'N 73°52.947'W). Use caution on this descent.

SOUTHSIDE TRAIL

Most hikers who go back and forth along the Phelps Trail never realize that there is a section of Johns Brook between The Garden and the Interior Outpost Ranger Station that is host to a number of small-to-medium-sized waterfalls—in fact, over eight of them on Johns Brook in addition to larger ones that have formed on two of its tributaries. The Phelps Trail simply stays too far away from Johns Brook for anyone to see the river. At best, hikers may hear it off in the distance once in a while. To view these rarely seen waterfalls, you must cross over to the red-blazed Southside Trail that parallels the Phelps Trail on the opposite side of Johns Brook.

There are three reasons why the Southside Trail is not heavily used, even though it also takes you to the Interior Outpost Ranger Station/Johns Brook Lodge junction just like the Phelps Trail does. First, most hikers are motivated to get into the heart of the High Peaks Region as quickly as possible. The well-worn, dependable Phelps Trail is a sure way to accomplish that goal. Second, a sign has been posted at the Southside Trail junction stating that the trail is eroded in sections and no longer being maintained. Many are discouraged by this message and go no farther. Third, even during most of its history when the Southside Trail was being actively maintained, crossing Johns Brook could prove enormously difficult. There is no bridge spanning the stream near The Garden end of the trail, and hikers must resort to rock-hopping to ford the stream. This is not so easily done during the spring and early summer.

I have hiked the Southside Trail in mid-September. At that time of year, Johns Brook is easy to cross (barring a prolonged downpour of rain), consisting more of stones than water. Once you get to the opposite side of the stream, the Southside Trail is in surprisingly good condition. It is an old, well-worn tote road that is very easy to follow. Along the way you must cross four tributaries coming into Johns Brook between The Garden and the Interior Outpost Ranger Station. While crossing these may be problematic in the early spring, they can be easily forded after the frenzy of spring's snowmelt has ended (unless, once again, a prolonged downpour of water has occurred). Only one section of the trail forces you to walk along the streambed for a short distance, but you shouldn't have any trouble managing this unless the brook is running high.

The paradox, of course, is that when the water level is low enough for Johns Brook to be easily forded, the waterfalls on the brook are not very robust; and when waterfalls are at their peak, crossing Johns Brook can be both difficult and dangerous. I suspect that the prime time to take the Southside Trail is when Johns Brook is water-filled but manageable to cross with water shoes and a walking stick. This is probably in the early summer or mid-fall.

For those who wish to visit these waterfalls at the height of their power, however, there is a way to do it. Instead of crossing over to the Southside Trail, follow the Phelps Trail all the way to the junction at the Interior Outpost Ranger Station, cross over the suspension bridge by the Ranger Station, and then follow the red-blazed Southside Trail northeast. When you come to Wolf Jaw Brook, you will be able to see the massive waterfall on this tributary and, if conditions are right, ford Wolf Jaw Brook downstream from the base of the fall and continue northeast on the Southside Trail.

Cascade on Johns Brook.

The hike at a glance: Let's take a moment to review what the Southside Trail consists of. Over the course of a 3.0 mile trek, you will be walking close to Johns Brook, often right next to it. However, in order to complete the hike, you will have to cross over four tributaries. These are, proceeding northeast to southwest: Rooster Comb Brook; Rock Cut Brook; Bennies Brook; and Wolf Jaw Brook. All are fairly close together. From Rooster Comb Brook to Rock

Cut Brook is a distance of ~1.1 miles; from Rock Cut Brook to Bennies Brook is ~0.5 mile; and from Bennies Brook to Wolf Jaw Brook is ~0.4 mile. Except in the spring, it should be fairly easy to ford the streams.

Photographs of small falls on Johns Brook can be seen in Eliot Porter's *Forever Wild: The Adirondacks* (1966), Bill Healy's *The Adirondacks: A Special World* (1986), and Gary A. Randorf's *The Adirondacks: Wild Island of Hope* (2002).

Another small cascade on Johns Brook.

The hike in detail: From the Garden, hike south along the yellow-blazed Phelps Trail for 0.5 mile. At a junction (44°11.026′N 73°49.333′W), turn left and follow the now-no-longer-maintained, red-blazed Southside Trail southeast for nearly 0.2 mile to Johns Brook. You will come out at a high embankment overlooking the stream, from where quite of bit of erosion is evident. Make your way down to the river and then ford it as best as you can. Water shoes and a walking stick for balance are always helpful when there is moderate flow. If too much water is flowing, turn back.

Once you reach the opposite shoreline, climb up the embankment and begin heading southwest. You will quickly pick up the continuation of the Southside Trail, which is a well-worn old tote road. For most of the hike, the trail is free of obstacles.

Within 0.2 mile you will cross over Rooster Comb Brook. I bushwhacked up Rooster Comb Brook for a short distance, but never saw any cascades. I did, however, come upon an upper road/trail where a log bridge crosses the stream.

Back on the Southside Trail, continue hiking southwest after crossing Rooster Comb Brook. Within 0.5 mile you will come to a fork. Bear right, taking the lower trail, which leads within 0.2 mile to an expansive area of bedrock containing a small cascade that flows into a 6–8-foot-wide pool (44°10.318′N 73°50.268′W). It is a very picturesque spot. As will be true for all of the sites described, the cascades and streambed will appear much different depending upon the time of year you are visiting.

The path that you are following quickly tapers off. Scamper down to Johns Brook and continue upriver along the rocky streambed for 300–400 feet. The bank to your left is so high and vertical that there is no option but to continue straight ahead. You will reach a huge pool where a small cascade enters through rocks (44°10.226′N 73°50.338′W). Like the previous cascade, the spot is scenic, but larger cascades await just up ahead.

First, however, you must climb up to the top of the bank since there is no way to follow the streambed any farther without getting seriously wet. Walk 50 feet back from the pool and climb up an obvious but steep path that leads to the top. You have now returned to where the trail paralleling Johns Brook continues.

Very quickly you will come to an 8-foot-high elongated cascade. In the springtime this cascade looks entirely different as more exposed bedrock adds to the waterfall.

One hundred feet farther upstream is a small cascade created by boulders that lies directly above a long underwater slab of bedrock. Resting next to the cascade is an enormous boulder. (I say "resting" because even large rocks are pushed along by the stream during times of high water.)

You now come to the first of the large waterfalls, formed on Rock Cut Brook, a tributary to Johns Brook. To see it, follow the trail down into a tiny gorge.

68. ROCK CUT BROOK FALLS

No

This is not just one waterfall, but several enjoined cascades (44°10.166'N 73°50.389'W). The lowermost one, directly below where the trail crosses, is a 4-foot-high slab cascade. Just upstream from the trail crossing is another 4-foot-high cascade. Then, 25 feet farther upstream is a 20-foot-high cascade partly filled with rocks. Still higher up is a 20–25-foot-high cascade where a large boulder tilted at a forty-five-degree angle lies across the middle of the cascade. Very unusual.

Most of this can be seen from the trail crossing, but to get a really good look at the upper cascade, hike up along the north side of the stream to the top of the lower cascade. At the time of year I was doing the hike, Rock Cut Brook was easy to cross. It may be more difficult to ford in the early spring.

Rock Cut Brook rises from the west shoulder of Hedgehog Mtn. (3,389'). Its name comes from the manner by which it has cut a narrow chasm into the bedrock

69. FUNNEL FALLS

No

In another 0.1 mile a spur path to your right leads down to a pretty, 8-foot-high waterslide that glides into a pool of water.

I am calling this cascade "Funnel Falls" (44°10.109'N 73°50.550'W) because of how the stream is funneled into the waterslide. A 2-foot-high cascade lies just above it. Many potholes are worn into the surrounding bedrock.

Return to the main trail.

70. TENDERFOOT POOL FALLS

In another 0.05 mile a steep spur path to your right leads down again to Johns Brook, this time to a 5-foot-high cascade that flows into a pool. From here the stream glides down a 1-foot fall into an even larger pool (44°10.086′N 73°50.616′W). Perhaps this is the section of the stream called Tenderfoot Pool Falls, or just Tenderfoot Falls as Poncho Doll refers to it in his 2002 book *Day Trips with a Splash: Northeastern Swimming Holes*. This whole area, composed mainly of solid bedrock, is completely different from the Johns Brook you initially crossed, where the streambed consisted of stones and rocks.

Return to the main trail. Immediately, you will come to a fork. Bear right, following the lower trail. In less than 0.05 mile you will reach a huge section of exposed bedrock on Johns Brook (44°10.023′N 73°50.701′W). When I visited this area, the stream was running to one side, producing a 2–3-foot-high cascade that faced the trail. Above it, a 6-foot-high waterfall loomed. During times of high water this entire section of bedrock becomes fully engaged, turning into one large, stream-wide cascade.

In a couple of hundred feet, the lower path reconnects with the upper one. Continue south. In less than 0.05 mile you will see to your right a long stretch of downed trees and debris—outwash from Bennies Brook, which is just up ahead. On the day of my visit, Bennies Brook was easy to cross, but I would imagine that the stream can be difficult to ford during early spring when the brook turns into a wide, raging torrent.

As soon as you cross Bennies Brook, turn right, following a path downhill for 50 feet. Then bear left onto a lower road/path that parallels Johns Brook, the goal being to always stay as close to Johns Brook as possible. You will immediately cross over a little rivulet that produces a couple of tiny, trailside cascades in the spring. Walk 50 feet farther south along the trail, then veer right and bushwhack over toward Johns Brook and down a 20-foot-high embankment to reach the streambed. The bushwhack, though short, can be a bit challenging.

71. JOHNS BROOK LOWER FLUME

You have now reached Johns Brook Lower Flume (approximately 44°09.897'N 73°50.804'W), the name I have given it in order to differentiate it from the flume/chasm near the Interior Outpost Ranger Station. If you have followed the directions as given, you will be standing at the bottom of an impressive, 150-foot-long flume with 8–10-foot-high walls. A 4-foot-high, boulder-chocked cascade comes in on the side of the flume. Water also enters from the end nearest you. One can only imagine what this flume must look like in mid-spring when a huge section of the flume's west wall turns into one long, broad cascade.

Johns Brook Lower Flume.

Climb back up to the trail. In less than another 0.1 mile a side path/road leads down past an old cabin and Adirondack-style lean-to to a huge area of scoured bedrock with many potholes. During high water a 10-foot-high cascade forms here. On the day of my visit, water was being funneled through a channel on the far side of the stream.

Although there were no posted signs, this camp is definitely on private property. A sign in the lean-to says "No camping." Give the camp a wide berth and focus strictly on the waterfall.

Return to the main trail. In less than 0.05 mile farther, a second spur path/road leads downhill to the river. Bushwhack over to Johns Brook and walk downstream for 100 feet to another area of exposed bedrock where a 5-foot-high cascade near the west bank has been produced. During times of high water the entire bedrock turns into one stream-wide cascade.

72. WOLF JAW BROOK FALLS

Go back to the main path, continue south, and you will immediately come to Wolf Jaw Brook. Although not apparent at first, just 75 feet upstream is massive Wolf Jaw Brook Falls (44°09.796'N 73°50.940'W) whose height I would estimate to be well over 150 feet. It is so massive that the waterfall cannot be seen in its 0.1-mile-long entirety from any one spot along its banks. In this respect Wolf Jaw Brook Falls reminds me of Glendale Falls in Massachusetts, a waterfall that I wrote about in *Berkshire Region Waterfall Guide*. The entirety of these falls can only be seen in sections. Wolf Jaw Brook Falls is probably best described as an enormous waterslide etched into a mountainside where everything trying to cling to the bedrock has been stripped away.

The name "Wolf Jaw(s)" is attributed to the famous artist Alexander Wyant, who came up with the name in 1869 while on Noonmark Mtn. painting what to him, off in the distance between two peaks, resembled a wolf's jaw.

Wolf Jaw Brook is a small-to-medium-sized tributary to Johns Brook that rises from the west shoulder of Lower Wolf Jaw (4,175'). It has the power of a mountain behind it to create significant water flow at times.

When you arrive at Wolf Jaw Brook, you will see a 4-foot-high cascade just upstream from the brook's confluence with Johns Brook. But this is just a tease. Cross over Wolf Jaw Brook and then bushwhack up past the small cascade. You will immediately come to an upper tote road. Turn left and scamper down to the base of Wolf Jaw Brook Falls. A former, wide bridge once crossed over the stream here, but it was demolished some time ago during one of Wolf Jaw Brook's rampages. Standing at the bottom and looking up, the medium-sized cascade that you see is only a small part of the entire waterfall. In fact, it wasn't until I crossed back over to the north side of the stream and began bushwhacking uphill that I realized just how massive this waterfall is. It just keeps going up and up.

A small part of Wolf Jaw Brook Falls.

From the base of the waterfall, follow the trail southwest for 0.1 mile to the junction with the trail to Upper and Lower Wolf Jaw that was cut in 1927 by the Adirondack Mountain Club. If you wish to see sections of Wolf Jaw Brook Falls without bushwhacking, follow the red-blazed Wolf Jaws Trail uphill for 0.1 mile. It conveniently passes by the top of the waterfall (44°09.615'N 73°50.899'W), allowing you to step off the trail for views. Realize that you will not be able to see the entire waterfall from the top. However, you will get a clear view looking west toward the Big Slide Mtn. area.

Before reaching the top of the waterfall, you can also follow a spur path to your left that leads to a midsection view of the waterfall. This is well worth doing.

From the Wolf Jaws Trail junction, continue south on the main trail for another 0.2 mile to reach the suspension bridge near the Interior Outpost Ranger Station and its marvelous chasm with waterfalls (see "Upper Johns Brook Flume & Falls").

Bonus Waterfall—Back at the kiosk at The Garden, instead of taking the Phelps Trail southwest, hike northwest on the blue-blazed Big Slide Mtn./Brothers Trail, proceeding uphill. At 0.2 mile turn right onto the red-blazed Porter Mtn./Little Porter Mtn. Trail and head north, going downhill for 0.1 mile into a picturesque gorge. At the footbridge crossing over Slide Brook (a medium-sized stream that rises from the northeast shoulder of Big Slide Mtn.), take note of a 4-foot-high, boulder-created cascade just upstream from the footbridge. It's what I call a "faux cascade," meaning that the cascade is created either by boulders or objects lying on the streambed and not by the streambed itself. The area is very wild and scenic.

VII: FROM KEENE TO KEESEVILLE

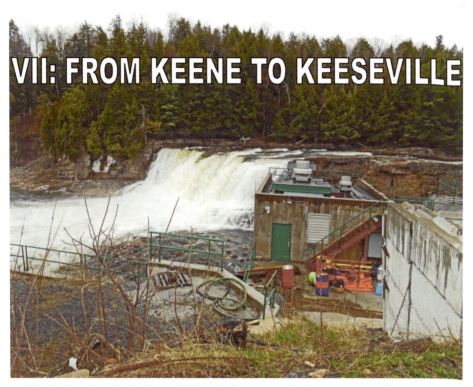

The hydroelectric station at Alice Falls serves to illustrate that the Ausable River is a working river.

The East Branch of the Ausable River continues north from Keene, becoming relatively docile after bouncing over Hulls Falls and through a long gorge outside of Keene. Once it reaches Jay, a series of small cascades and waterslides are encountered. After Jay, the river again becomes fairly placid as it meanders along a relatively flat landscape encased in a deep valley. At Ausable Forks the East Branch combines forces with the West Branch to form the powerful Ausable River. It is at Keeseville where the river begins to change dramatically, producing one waterfall after another: Anderson Falls in Keeseville; Alice Falls at Alice Falls Hydro; a dammed cascade upstream from Rainbow Falls; Rainbow Falls at Ausable Chasm; and, finally, Horseshoe Falls at Ausable Chasm.

There is much to see as we follow the East Branch north.

73. STYLES BROOK FALLS

Styles Brook Falls, sometimes spelled "Stiles," is a large, block-shaped waterfall about 40–50 feet high that is formed on Styles Brook, a medium-sized stream that rises from the Jay Mountain Ridge and flows into the East Branch of the Ausable River. Several tenths of a mile downstream from Styles Brook Falls is the Styles Brook Falls Bridge (erected in 1931 and rehabilitated in 1990) that Route 9N crosses over. Upstream from the fall is an impressive, 0.2-mile-long gorge rippling with many small cascades and rapids.

Topographical software suggests that the waterfall is located at a GPS coordinate of around 44°17.792'N 73°46.703'W.

During the nineteenth century a sawmill operated near the base of the waterfall. Its position is shown on the Burr Map of 1829 in a flat area now occupied by several camps.

According to one source, a seven-mile-long water chute (called a "log slip") once sent volleys of logs down the Styles Brook Valley from above the Glen. The chute, consisting of a water-filled wooden trough, was constructed by the W. & R. Rogers Company. A photo of this chute (or one very similar to it) can be seen in Maitland C. Desormo's *The Heydays of the Adirondacks*. Similar log slips operated on Mount Colden above Avalanche Pass, below Big Slide Mountain, and at a number of other sites.

It is believed that some of the settlers who ventured west from Lewis used the Styles Brook–Sprucemill Brook notch to make their way over the mountains. They likely passed by Styles Brook Falls in doing so.

Styles Brook Falls. Postcard c. 1920.

A photograph of the waterfall can be seen in Derek Doeffinger & Keith Boas's *Waterfalls of the Adirondacks and Catskills* (2000), Nathan Farb's *100 Views of the Adirondacks* (1989), and Gary A. Randorf's *The Adirondacks: Wild Island of Hope* (2002).

To get there: From the hamlet of Keene (junction of Routes 73 & 9N North), head north on Route 9N North, paralleling the East Branch of the Ausable River to your left. After 3.0 miles you will cross over a medium-sized stream coming in on the right; this is Styles Brook. Immediately turn right onto Styles Brook Road (Route 52). The road climbs steeply, paralleling Styles Brook, which is at some distance to the right and cannot be glimpsed through the woods. The waterfall is only 0.3–0.4 mile upstream from the bridge, but the land by the waterfall is posted, so access without permission is illegal.

It is unclear whether access to the upper sections of the gorge is permitted, but it may be possible to legally reach the gorge and smaller cascades by following a power line from Styles Brook Road south into the woods (44°17.893'N 73°46.525'W). The power line is 0.6 mile up Styles Brook Road from Route 9N. The power line is perpendicular to a well-worn path that follows along the top of the gorge in both directions.

Interestingly, farther uphill on Styles Brook Road is a tiny section of state land where Styles Brook can be legally accessed, but it is too far up from the gorge to be of any benefit in viewing the gorge.

74. JAY FALLS

Jay Falls (44°22.371'N 73°43.567'W) consist of a series of cascades and waterslides dropping about 20 feet over a large area of exposed bedrock. They represents the last falls you will come across on the East Branch of the Ausable River before it joins with the West Branch at Ausable Forks. In *Adirondack Canoe Waters: North Flow* (1987), Paul Jamieson & Donald Morris write that "Jay Falls is a potentially dangerous Class V–VI drop in several sections," meaning that the waters here are particularly treacherous for paddlers attempting a whitewater run.

A number of mills, including a forge, brickyard, wheelwright shop, gristmill, and blacksmith shop, once operated by or near the falls. Foundation remnants can be seen along the south bank.

A photograph of the falls with swimmers and rock-hoppers can be seen in "Holy Waters" by Ben Stechschulte in the August 2005 *Adirondack Life*. A very artistic photograph of Jay Falls can be seen in *Waterfalls of New York State* by Edward M. Smathers, Scott A. Ensminger, & David J. Schryver, as well as a 2004 shot looking down from the top of the falls toward the covered bridge in Den Linnehan's *Adirondack Splendor*.

The cascades and historic covered bridge (downstream from the falls) make for a winning combination. Photograph by John Haywood.

The 1857 Howe Truss Span Bridge, which replaced an earlier bridge washed away in the "freshet of 1856," is located just downstream from the cascades. It opened in October 2007 as a pedestrian footbridge and now serves as the centerpiece of a wonderful park overlooking the cascades, complete with historic markers and plaques. When Barbara and I visited Jay a number of years ago, the covered bridge was just lying patiently along the side of the road waiting to be rehabilitated. It must have been an epic project lifting that bridge off its abutments and then returning it years later, refurbished, to its former position.

Both the hamlets of Jay and Upper Jay (which you passed through on the way to Jay) are named after John Jay, former president of the Continental Congress, ambassador to Spain, first Chief Justice of the United States, and New York State governor. Earlier the town of Jay was called Mallory's Bush, named after Nathaniel Mallory, the town's earliest settler.

To get there: From Keene (junction of Route 73 & 9N North) drive north on Route 9N North for ~9.6 miles. At the center of Jay, turn right onto John Fountain Road (Route 22) and drive southeast for less than 0.2 mile. Turn right onto a short dead-end road and park by the north end of the bridge (44°22.405'N 73°43.508'W).

AUSABLE FORKS

After leaving Jay behind and heading north for ~5.5 miles, the East Branch of the Ausable River reaches Ausable Forks, a former iron manufacturing village that was settled in 1824. The town has been called the "Gateway to the Olympics and High Peaks" because of its proximity to Lake Placid, Whiteface Mountain, and Keene.

Indications of the town's industrial past are evident in its street names—Forge Street, Rolling Mill Hill Road, and Crusher Road (named after a powerful crusher that separated rock from ore).

At Ausable Forks the East Branch joins with the West Branch to form the Ausable River (44°26.440'N 73°40.467'W). There are no waterfalls in Ausable Forks. Continue on Route 9N to Keeseville paralleling the Ausable River, now to your right.

75. ANDERSON FALLS

Anderson Falls (44°30.276'N 73°28.908'W) is a 15-foot-high, waterslide-like cascade formed on the Ausable River and the first waterfall encountered after the two branches of the Ausable River join forces. According to Paul Jamieson & Donald Morris in *Adirondack Canoe Waters: North Flow*, "This waterfall drops 18 feet in 50 yards." The waterfall is named after John Anderson, an early settler and lumberman. For a time the village was also called Anderson

Falls. In 1812 the town's name changed to Keeseville in honor of Richard Keese, a prominent local businessman, and it has kept that name ever since.

In 1858 a group of Boston luminaries including William Stillman, Ralph Waldo Emerson, Louis Agassiz, James Russell Lowell, and six other men of distinction passed through the village and probably saw Anderson Falls while on their way to Follensby Pond to set up what would later become known as the Philosophers' Camp.

In the early 1800s a settler named Robert Hoyle established a store near the waterfall, which he called "The Long Chute."

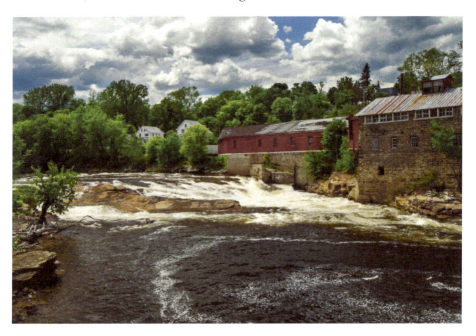

Anderson Falls was named after John Anderson, an early settler and lumberman. Photograph by John Haywood.

Downstream from the fall is the historic 110-foot-long keystone Arch Bridge designed by Solomon Townsend. It was built in 1841–1843 and is reputed to be the second-largest single-arch stone span in the United States. The village's Main Street (Route 22) crosses directly over it. Near the stone arch bridge on the west bank is the former 1852 Ausable Chasm Horsenail Works. Next to it is a long red-colored factory building. On the opposite bank public walks have replaced the long-gone plaster mill, gristmill, twine mill, factory machine shop, and furniture factory.

Just a short distance upstream from the cascade is a 240-foot-long pedestrian suspension (swinging) bridge that was built in 1888 by the Berlin

Iron Company of East Berlin, Connecticut. It's fun to walk across while looking downriver toward the waterfall and stone arch bridge.

A dam near the top of the fall was washed away in the 1970s.

To get there: From Ausable Forks proceed northeast on Route 9N for ~10.0 miles until you pass under Exit 34 of the Adirondack Northway (I-87). From here continue northeast on Route 9N (Pleasant Street) for 1.1 miles. When you come to Main Street, turn right and head east for less than 0.2 mile. Then turn right onto Front Street and proceed south for 0.1 mile. At Clinton Street turn right, go 100 feet, and then bear right onto Mill Street and park to your right (44°30.219'N 73°28.902'W).

From Exit 33 of the Adirondack Northway (I-87), head north on Routes 9/22 for 4.2 miles. Turn left onto Clinton Street and drive west for 100 feet. You will come to Mill Street. Park to your right.

From the parking area, walk down Mill Street for less than 100 feet to a park that overlooks the falls. The park, including a pavilion and benches, was created in the 1980s and occupies the spot where the R. Prescott & Sons Furniture Factory once stood until it burned down in 1966.

It is also possible to descend to the level of the streambed following a stone path between two walls.

The park parallels the river all the way down to the stone Arch Bridge, just downstream from the cascade.

76. INDIAN FALLS

Indian Falls is a 6–8-foot-high waterfall formed on the Ausable River between Anderson Falls and Alice Falls, approximately 1,100 feet upriver from Alice Falls. According to Paul Jamieson & Donald Morris in *Adirondack Canoe Waters: North Flow* (1987), the fall is rated as a Class III–IV drop by whitewater paddlers.

The waterfall was once the site of the Ausable Chasm Horsenail Lower Works, a match factory, and a rolling mill. Foundations of the lower works and rolling mill are still visible.

To get there: From Keeseville (junction of Routes 9 & 9N) drive north on Route 9 for ~0.8 mile. After you pass by the motels along Route 9, you will start to go slightly uphill. Just before the top of the hill, turn right into a pull-off. Follow a well-worn trail used by fishermen that leads to the fall (44°30.984'N 73°28.082'W).

Indian Falls was once the site of considerable industrial activity. Photograph by Heidi Mosley.

77. ALICE FALLS

Alice Falls (44°31.144'N 73°27.861'W) is a 35–40-foot-high dammed waterfall formed on the Ausable River ~1.7 miles downriver from Anderson Falls and less than 0.5 mile upriver from Rainbow Falls. Don't let the presence of the dam or the fact that the fall is used for hydroelectric power discourage you from stopping to see this waterfall. Alice Falls is well worth the visit, and the plant manager, if he is in attendance, will patiently answer any questions you may have about the waterfall and the power station.

Alice Falls is an artificial waterfall. The original waterfall was closer to the opposite bank until the Alice Falls Hydroelectric Project was constructed and the course of the river altered to its present position. Another interesting fact about Alice Falls is that one part of it lies in Essex County, and the other part (the one farthest away) in Clinton County, which makes it a two-county waterfall.

Alice Falls was first called Little Falls, no doubt to distinguish it from larger Rainbow Falls a short distance downriver. The origin of the name Alice Falls, however, is unknown.

Forty-foot-high Alice Falls is only a short distance upriver from Ausable Chasm.

In *Exploring the Adirondack Mountains 100 Years Ago*—a series of nineteenth-century articles compiled and published by Stuart D. Ludlum in 1976—Alice Falls and Rainbow Falls are described, but incorrectly named. Alice Falls is called Birmingham Falls, and Rainbow Falls is called Great Falls. Keep that in mind as you read the following passage: "At the distance of a mile or so from Keeseville is Birmingham Falls [Alice Falls], where the Ausable descends about thirty feet into a semicircular basin of great beauty; a mile farther down are the Great Falls [Rainbow Falls], one hundred and fifty feet high, surrounded by the wildest scenery." In this description Alice Falls is given a reasonable height estimate, but Rainbow Falls, at 150 feet, is way off the charts. Accompanying the article is a line drawing by nineteenth-century illustrator Harry Fenn where, once again, the height of the waterfall is exaggerated, only this time visually. It is not an exaggeration to say, however, that Alice Falls is a beautiful waterfall at any time of the year. A photograph of the Alice Falls Pulp Mill, c. 1890, can be seen in Steven Engelhart's 1991 book, *Crossing the River: Historic Bridges of the Ausable River*.

To get there: From Keeseville (junction of Routes 9 & 9N) drive northeast on Route 9 for over 1.0 mile. Turn right at the sign for Alice Falls Hydro Park to your left before you reach the manager's rustic log cabin (44°31.197'N 73°27.891'W). There are great views of the fall from the park.

78. DAMMED WATERFALL

This unnamed, 15-foot-high waterfall, layered with ledges and topped by a large dam, is formed on the Ausable River. There are some who would contend that the waterfall is part of Rainbow Falls, which lies immediately downriver. We consider this waterfall to be a separate entity all to itself.

To get there:
Access #1: Continue northeast on Route 9 from Alice Falls for over 0.1 mile and turn right onto Old State Road (now a dead-end street). After heading east for 0.2 mile, park to your left at the end of the road by a bridge that, at one time, was the main route north over the Ausable River, but which has been closed to traffic for many years (44°31.379'N 73°27.648'W). There are excellent views of the fall from the bridge.

Access #2: Continuing north on Route 9 from Alice Falls, cross over the large bridge spanning the Ausable River at Ausable Chasm and turn right onto Route 373. Drive southeast for 0.2 mile. Turn sharply right onto Mace Chasm Road (Route 71) and head south for 0.3 mile. Park at the east end of the closed bridge (44°31.426'N 73°27.575'W). Look upstream for close-up views of the dammed fall and downstream for views over the top of Rainbow Falls.

Dammed waterfall above Rainbow Falls. Postcard c. 1940.

AUSABLE CHASM

Ausable Chasm is one of the great natural wonders of the eastern United States and remains the longest continuously operating commercial attraction in America. It has been called both "Yosemite in miniature" and the "Grand Canyon of the East."

Ausable Chasm is nearly 2.0 miles long, with over 0.5 mile of it traversable by boat. Stairways and walkways enable visitors to descend into the interior of the Chasm along the Inner Sanctum Trail. In addition to waterfalls there are interesting rock formations in the Chasm, including Elephant Head, Jacob's Well, Hyde's Cave, the Devil's Oven, and a number of other named features. For those who like to hike, rim trails follow along both sides of the Chasm, providing outstanding overlooks. There are even two dry chasms, long ago abandoned by the river, that parallel Ausable Chasm and can be explored at length.

It's hard to look at a scenic wonder like Ausable Chasm and not wonder just how such a marvelous creation of nature came about. In 1965 Dr. Charles E. Resser wrote an article for the *York State Traditions* magazine listing four necessary conditions for a canyon like Ausable Chasm to form. First, there must be a swift-moving river racing over flat-lying bedrock. Tilted bedrock simply won't work. Second, the bedrock must be composed of block-like layers that can be easily eroded, leaving behind vertical walls. Third, the walls have to be strong enough to resist chemical and mechanical erosion, enabling them to remain vertical for long spans of time. Lastly, eons must have already passed for the chasm to form, but not so many eons that its walls have matured to form a V-shaped gorge.

Ausable Chasm will always have a special place in my heart. Between 2014 and 2015, I spent countless hours working with photographer John Haywood and Ausable Chasm historian and guide Sean Reines to write the definitive guidebook to Ausable Chasm that was published in 2015. It's called *Ausable Chasm in Pictures and Story* and is a must-read booklet if you are going to visit the Chasm. It contains not only geological and historical facts about the Chasm, but fascinating pictures as well.

Ausable Chasm is located at 2144 Route 9, Ausable Chasm, NY 12911 and can be contacted at (518) 834-7454 or through their Web site, ausablechasm.com.

To get there: From Keeseville (junction of Routes 9 & 9N) drive north on Route 9 for 1.5 miles. As soon as you cross over the Ausable Chasm Bridge, turn left into the entrance for Ausable Chasm and park near the Visitor Center (44°31.511′N 73°27.728′W).

From the Adirondack Northway going south, take Exit 35 for Peru/Port Kent and turn left onto Bear Swamp Road (Route 442). After driving east for 2.9 miles, turn right onto Route 9 and head south for 3.7 miles. Then turn right into the entrance for Ausable Chasm.

If you want to get close up to Rainbow Falls and Horseshoe Falls, be sure to sign up for the Adventure Trail Tour that takes you down into the section where these two waterfalls can be seen more intimately.

79. RAINBOW FALLS

Rainbow Falls (44°31.420′N 73°27.625′W) is a 70-foot-high waterfall formed on the Ausable River only a few miles upstream from the river's terminus at Ausable Point. Without a doubt Rainbow Falls is one of the Adirondacks' most accessible and photographed waterfalls, as it is located at the head of

Ausable Chasm just upriver from a high bridge overlooking it. Although a considerable volume of water goes over the top of Rainbow Falls continuously, significant amounts are also bled off for power generation and then returned to the Chasm, producing a secondary waterfall at a right angle to the main fall. Some writers have mistaken this side fall for Rainbow Falls, perhaps because it captures more than its share of rainbow-producing sunlight.

Rainbow Falls has been known by other names. Early on it was called Adgate Falls after Matthew Adgate, who erected a sawmill and gristmill at the fall. Then, during the river's heavy industrial era, it was known as Birmingham Falls in honor of the industrial town of Birmingham, England. Its present name comes from the rainbow produced by tiny droplets of spray turned into little prisms by the sunlight.

In *The Adirondacks Illustrated*, Seneca Ray Stoddard describes descending to the bottom of the Chasm to where "the spray from the cataract which, divided in the center, falls in almost unbroken sheets a distance of seventy feet." Stoddard's estimate of the waterfall's height is pretty accurate, as is Duane Hamilton Hurd's in *History of Clinton and Franklin Counties, New York*, who writes, "Birmingham Falls [Rainbow Falls], at the opening of the chasm, has a perpendicular descent of seventy feet, and extends diagonally across the river, almost directly facing the western shore."

In *The Military and Civil History of the County of Essex, New York*, Winslow C. Watson writes: "Foaming and surging from this point, over a rocky bed, until it reaches the village of Birmingham [Ausable Chasm], it then abruptly bursts into a dark, deep chasm of sixty feet. A bridge, with one abutment sitting upon a rock that divides the stream, crosses the river at the head of this fall. This bridge [located near the top of Rainbow Falls] is perpetually enveloped in a thick cloud of spray and mist. In winter, the frost work encrusts the rock and trees, with the most gorgeous fabrics, myriads of columns and arches, and icy diamonds and stalactites glitter in the sunbeams. In the sunshine a brilliant rainbow spreads its radiant arc over this deep abyss."

Regrettably, Rainbow Falls hasn't always been fully appreciated for its awe-inspiring beauty. In 1876, six years after Ausable Chasm opened as a commercial venture, the Ausable Chasm Horsenail Works erected part of its factory at the base of Rainbow Falls along the north side of the Chasm. This business venture lasted until 1910. Look closely from the Route 9 bridge spanning the Chasm and you will see the surprisingly intact stone ruins of the horse nail factory's wheelhouse, an 1880s photograph of which can be seen in Jeffrey L. Horrell's 1999 book, *Seneca Ray Stoddard: Transforming the Adirondack*

Wilderness in Text and Image. Considering how much water rushes through the Chasm like a battering ram during spring's freshets, the fact that this structure is still standing is nothing short of a miracle.

Seventy-foot-high Rainbow Falls lies at the head of Ausable Chasm. Photograph by John Haywood.

Up until 1971 the color of the water going over the fall was either a muddy pink or a murky green, typically prompting visitors to ask the Chasm guides why the water was colored. The answer had to do with a paper mill on the West Branch of the Ausable River upstream from Ausable Forks that used different colored dyes. Some days the dye was pink, other days green. Eventually the mill closed, mainly because it couldn't comply with new anti-pollution regulations. The water is now perfectly clear.

For those interested in statistics, a retired actuary and statistician named Richard H. Beisel, Jr. developed a waterfall rating system and subsequently published a book in 2006 called the *International Waterfall Classification System*. On a scale of 1–10, he gave Rainbow Falls (with an average discharge of 11.3m^3/sec) a Beisel Waterfall Rating of 3.2. That's a pretty good number. In case you're wondering, there are only two waterfalls in the world that received a Beisel Rating of 10.0. The first one is Niagara Falls. The second waterfall and, of the two, the only one unaltered by man, is Boyoma Falls, aka Stanley Falls, in the Democratic Republic of the Congo.

To get there: As soon as you cross over the Route 9 bridge spanning the Ausable River and Ausable Chasm, turn left, follow the road down and around, and park close to the Visitor Center. Rainbow Falls can be seen by walking up to the Route 9 bridge or by walking past the Visitor Center, staying close to the rim of the Chasm, for a lateral view. No one, however, should miss the opportunity of going into Ausable Chasm for a guided tour, which includes not only amazing close-up views of Rainbow Falls but of many other natural features of the Chasm as well.

80. HORSESHOE FALLS

Horseshoe Falls, downstream from Rainbow Falls, adds immeasurably to the scenic view of Ausable Chasm. Photograph by John Haywood.

Horseshoe Falls (44°31.479'N 73°27.770'W) is a 6–8-foot-high waterfall formed on the Ausable River in Ausable Chasm just a short distance downriver from Rainbow Falls. The name Horseshoe Falls comes from the shape of the waterfall, but it's also possible that it may derive in part from the horse nail factory that once operated just upstream from the fall.

To get there: Follow the same directions given for Rainbow Falls. Horseshoe Falls can be easily seen from the Route 9 bridge spanning the Chasm or, better yet, from the interior of the Chasm on a guided tour.

81. TINY CASCADES AND RAPIDS

Downstream from Rainbow Falls, Horseshoe Falls, and the Route 9 bridge, the river changes dramatically. The walls narrow and the river accelerates, becoming more violent and churned up (44°31.620'N 73°27.710'W). Rapids and small cascades are produced in places. While no one cascade or rapid is particularly deserving of attention, it is well worth your time to hike down on the Chasm's Inner Sanctum Trail to see for yourself the power of moving water.

Once beyond the rapids, the river quiets down as it passes through the Flume. Here, the walls contract to as little as thirteen feet from side to side, and the depth of the river increases to 60 feet even during times of low water flow. Almost half of the Chasm in the Flume remains perpetually underwater. You do not want to be caught in this section of the Chasm during times of high water flow, however, for the level of the river can rise to as much as 40 feet above the trail!

Hells Gate. Ausable Chasm. Postcard c. 1940.

From Ausable Chasm northeast to Lake Champlain, the waters are relatively placid. As Winslow C. Watson writes in his 1869 book, *The Military and Civil History of the County of Essex*, "No mill site occurs below Birmingham

[Ausable Chasm] upon the river, but the project exists of erecting large mills at the mouth of the Au Sable to be propelled by stream." To the best of my knowledge, this never came about.

Ausable Chasm is not the last word on Ausable River–created waterfalls, however. The Little Ausable River, just north of the Ausable River and east of Laphams Mills, has produced its own series of cascades, possibly known as Cooper's Falls, in a small gorge.

82. BUTTERMILK FALLS

Buttermilk Falls (a common name for a waterfall) is located on private property and can be accessed only with the permission of the owner.

The waterfall is reputedly 168 feet in height and formed on Little Trout Brook, a small stream that rises from the northwest shoulder of Skagerack Mtn. and flows into Lake Champlain south of Port Douglas. A photograph of the fall can be seen in *Around Keeseville: Images of America* by Kyle M. Page.

The waterfall is significant historically. During the late nineteenth and early twentieth century it was frequented by tourists, travelers, and fishermen from Port Douglas. The Hotel Douglas, in fact, used to arrange trips to the fall as part of their regular itinerary for guests. Whether guests ever routinely descended to the base of the waterfall is unknown, for the way down is not obvious. More likely, visitors hiked upstream from a lower point to the bottom of the cascade.

Pitchoff Mtn. from Old Military Road.

VIII: FROM KEENE TO HEART LAKE

From Keene (junction of Routes 73 & 9N North next to the presently abandoned Monty's Elm Tree Inn) there are two main routes that lead out of town. Route 9N North follows along the East Branch of the Ausable River and is already familiar to us for its waterfalls in Jay, Keeseville, and Ausable Chasm. Route 73 takes us in an entirely different direction, heading toward Lake Placid. This is the route that we will follow in Section VIII.

First, a few words about Keene. This tiny village prospered during the eighteenth century because of its strategic location along the Northwest Bay Trail and its close proximity to the Ausable's East Branch, where waterpower was abundant. By the end of the 1700s, the community was significant enough to show up on regional maps.

Up until the late 1850s, travelers journeying between Keene and North Elba/Lake Placid had to take the Mountain Road (part of the Old Military Road) north of Pitchoff Mountain. According to Mary MacKenzie in *The Plains of Abraham*, this was a deplorable road referred to by travelers as "six miles, six hours." Travel options changed in 1859 with the creation of the Cascade Pass Road that took travelers past two lakes of extraordinary beauty hugged by towering mountains on each side.

There are two relatively small falls to see before we head too far west on Route 73. One involves a 1.0-mile hike; the other can be partially glimpsed from the top of a bridge.

83. NICHOLS BROOK FALLS

Nichols Brook Falls consist of a series of small cascades formed on Nichols Brook. Although they are not large, they do take you to a wilderness area visited by few, and this should be compensation enough for your efforts.

The path that you will be following next to Nichols Brook, on your left, is the Old Military Road/Mountain Road, which dates back to the end of the eighteenth century when it was part of the Northwest Bay–Hopkinton Road. Until the creation of a road in 1858 (an early version of what would become part of Route 73) that went past the Cascade Lakes, the Old Military Road served as the only means of getting between Keene and North Elba and then to Lake Placid and beyond.

Although no one knows for certain, it's quite possible that Mary Brown—wife of the famous abolitionist, John Brown—chose the Old Military Road rather than the newly constructed Cascade Pass Road when bringing her husband's body back to North Elba. Sections of the old road are still traversable by car—three miles of highway on the Keene end (called Alstead Hill Road) and one mile of highway on the North Elba end (called Mountain Road). The 4.5 miles of road in between are closed and essentially impassable by vehicle. Since 1986 the Adirondack Ski Touring Council has assumed responsibility for maintaining the in-between section as a foot trail, and the route has become a part of the Jack Rabbit (Ski) Trail, named after a Norwegian, Herman "Jack Rabbit" Johannsen, who laid out part of the original trail between 1916 and 1928. Johannsen lived to be 111 years old, dying in his native country of Norway.

The Old Military Road was constructed by the equivalent of today's Army Corps of Engineers and aptly named, for it may have served to facilitate the movement of troops as well as providing an inroad for Revolutionary War veterans looking to settle on huge military tracts set aside in the Adirondacks by the New York State Legislature.

The trail to Nichols Brook Falls is slowly falling into disuse.

To get there: From Keene (junction of Route 73 & 9N North) drive north on Route 73 for 0.8 mile, then turn right onto Alstead Hill Road and head west for 2.9 miles. Along the way, at 0.4 mile, you will pass by the Bark Eater Inn (barkeater.com), which originally was a stagecoach stop at an old family farm. Park at the drivable end of the road (44°16.000'N 73°51.045'W), next to Adirondack Rock & River Guide Service—a lodge and guiding service at 616 Alstead Hill Lane, Keene 12942, (518) 576-2041.

Follow the trail from the parking area along the non-drivable part of the Old Military Road. It is wide and provides a fairly graded walk. At 0.8 mile a wooden bridge takes you over Nichols Brook. After another 0.1 mile, just before the trail reaches an enormous meadow, turn right at 44°15.661'N 73°52.043'W onto what was once a yellow-blazed path heading north. It would appear that the trail is no longer being maintained. The walk here is fairly short, no more than 0.1 mile in length. The cascades are encountered below where the path used to cross the stream. To see the falls in their entirety, you must be willing to bushwhack a bit along the bank of the brook.

84. CLIFFORD FALLS

No PRIVATE

Clifford Falls is formed on Clifford Brook, a small stream that rises from the southeast end of the Sentinel Range and flows into the East Branch of the Ausable River. The waterfall is tucked away near the end of a backcountry road that leads to a private residence. It lies directly below the 1895 Clifford Falls Bridge, making it hard to see because Clifford Brook turns at a right angle here. In the past, waterfallers would scramble down an embankment below the fall via a well-worn herd path for views of the cascade. This is no longer permitted. Brush and debris have been thrown across the top of the bank to block access, and "no trespassing" signs glare down at those who might have second thoughts about turning away.

You can still obtain views looking down from the bridge, but very little of the 4-foot-high waterfall can be glimpsed. Tantalizingly, the sound of falling waters clearly makes the waterfall's presence known.

Clifford Falls.

Small cascades and rapids can be viewed upstream from the bridge, and those are worth a look.

Photographs of sections of Clifford Falls can be seen in Nathan Farb's 1985 book *The Adirondacks* and George Wuerthner's 1988 book *The Adirondacks Forever Wild*.

Across the bridge, at the end of the road, is an 1835 farmhouse that has supported both maple sugaring and horse farming.

I've read that another small fall has formed on Clifford Brook a short distance farther downstream, but I have not been to it. It, too, is on private property and thus inaccessible.

To get there: From Keene (junction of Routes 73 & 9N North) drive north on Route 73 for 0.8 mile. Turn right onto Alstead Hill Road and head northwest for 1.0 mile. When you come to Bartlett Road, turn right and proceed north for 0.6 mile. Then turn left onto Clifford Falls Lane and go southwest for over 0.4 mile. Pull over to the side of the road and park before reaching the bridge (44°16.615′N 73°49.390′W).

85. CASCADE LAKE FALLS

Cascade Lake Falls (44°13.468′N 73°52.341′W), aka Cascade Falls, is a towering, 150–200-foot-high waterfall formed on a small seasonal stream called Cascade Brook that rises from the north shoulder of Cascade Mtn. (4,098′) and flows into the Cascade Lakes.

In Winslow C. Watson's 1869 book, *The Military and Civil History of the County of Essex*, the author writes, "In the bed of a little brook which leaps down the slide of the avalanche several hundred feet forming a succession of exquisite cascades, innumerable minerals sparkle and glow in every direction." The brook on the slope downstream from Cascade Lake Falls is a veritable rock garden for mineral collectors.

According to Phil Gallos in his 1972 book, *By Foot in the Adirondacks*, "The cascade of the Cascades is a waterfall of well over two hundred feet high, the base of which is located one hundred or so feet above the isthmus that divides the two lakes."

This is one waterfall that can be glimpsed from Route 73, but only in passing. To get a really good look, visitors need to turn onto the short road that leads down to an isthmus between the two Cascade Lakes.

A photograph of Cascade Lake Falls can be seen in Den Linnehan's 2004 book *Adirondack Splendor* and in James Kraus's *Adirondack Moments* (2009).

The Upper and Lower Cascade Lakes are formed in a narrow pass that lies between Pitchoff Mtn. (~3,600′) and Cascade Mtn. (4,098′). The upper lake is 0.5-mile long; the lower lake is ~1.0-mile long. The Cascade Pass fault extends for over ten miles from North Elba to the village of Keene. There are some who contend that the two lakes were once part of one larger lake until an avalanche in the early 1800s swept across the middle, dividing the lake into two bodies of waters, but this is unsubstantiated and probably apocryphal.

In an 1881 piece called "Adirondack Days" that appeared in *Harper's New Monthly Magazine*, the Cascade Lakes are described in the flowery, hyperbolic prose of that time period: "These ponds are twin sheets of water, which completely fill a long and narrow defile in the mountains. I will not compare them to great pearls fallen into a crevice of malachite, nor to mirrors in which the mountains, like Narcissus, study their own beauty, for I must confess that such comparisons always seem to me trivial and unworthy. No jewels or works of art ever equaled the strange, wild loveliness of these little lakes."

Cascade Lake Falls in mid-autumn.

The lake(s) was originally called Long Pond. Later the name changed to Edmund's Pond to honor a family in Keene, but this name also proved short-lived. Around 1878, Nicanor and Edna Miller built a summer hotel between the two lakes and rechristened the lakes "Cascades Lakes" to give proper recognition to the high waterfall that so prominently loomed above the valley floor. The hotel was a three-story, barn-like structure that could accommodate up to 50 guests. Undoubtedly the Millers promoted the notion that the hotel's elevation of 2,045 feet provided health benefits not found in the valley below, as was believed in those days.

In 1883 the Millers sold the hotel to Warren F. Weston and J. Henry Otis, who enlarged the structure to three times its initial size to accommodate the increasing number of tourists. In its prime the hotel offered a piazza, bowling alley, lawn tennis court, and a roller-skating rink to customers. It even had its own post office, called Cascadeville, which operated seasonally. An old photograph of the Cascade Lake House can be seen in the 1991 book, *Of the Summits, Of the Forests: ADK 46-R*, edited by Tim Tefft.

After 1905 the Cascade Lake House began to decline as tourism in the Adirondacks in general went into a downswing. In 1923 the property was sold to the Lake Placid Club, including 1,440 acres of land. The hotel was closed for good in 1929. In the 1930s the hotel, now abandoned, was seriously damaged when workers blasted rock on Route 73 while creating today's modern road. A photograph of Route 73 under construction can be seen in the 1999 book, *Two Adirondack Hamlets in History: Keene and Keene Valley*. It was taken by John Apperson in 1933; his name will come up again in the write-up for Indian Falls.

It wasn't long after that the hotel was completely demolished. Today little evidence of a large hotel remains, but if you look closely you can still see remnants here and there. Traces of the hotel's reservoir also still exist near the base of the falls. In 1951 the property was sold to the State of New York and has been "forever wild" since then.

To get there: From Keene (junction of Routes 73 & 9N North) drive northwest, then southwest on Route 73 for over 5.8 miles. Turn left onto a small, 0.05-mile-long road that leads down to the isthmus between the two Cascade Lakes. Park here (44°13.538'N 73°52.506'W).

From Lake Placid (junction of Routes 73 & 86) drive east on Route 73 for ~8.4 miles and turn right onto the short road that leads down to the isthmus.

After you park, take note of the stone wall ruins to your left at the end of the road. More hotel ruins can also be seen several hundred feet to your right near the south corner of Lower Cascade Lake. You will notice that the waters from Upper Cascade Lake flow into Lower Cascade Lake and from there make their way downhill, eventually flowing into the East Branch of the Ausable River in Keene.

After exploring the immediate area, follow the trail toward Cascade Lake Falls. The path starts off easy, but within 0.05 mile begins to increase in difficulty because of damage inflicted by the last tropical storms. Go to the right of a high mound of outwash and blowdown. From here continue uphill following a more recently made trail until you reach a flat area where decent views of the upper parts of Cascade Lake Falls can be obtained.

You can continue even farther upstream from this point, following a faint trail that goes off to the left steeply above the stream. After another hundred feet it will take you to a second look-out area.

Keep in mind, of course, that these directions can be altered at any time by severe storms.

Phil Gallos, in *By Foot in the Adirondacks*, describes the top of the falls, writing, "Here, high above the lakes, we find several pools linked by small waterfalls all set in a narrow glen at the brink of the great Cascade."

I've read that a trail once led up through a gully past the cascade on the way to the summit. The trail was created by the hotel owners in 1891 but was destroyed by fire in 1903. Miraculously, the hotel itself was spared during the conflagration.

Truly, the best view of the waterfall is actually from the parking area, where it can be seen in its entirety.

Cascade Lake Falls #2: There is also a second Cascade Lake Falls—a smaller, more seasonal version that drops into Upper Cascade Lake not far from the southwest corner of the lake. I'm giving it a GPS reading of 44°13.279'N 73°52.766'W, but this is at best a crude estimate. Bear in mind that Cascade Lake Falls #2 is on a narrow rivulet and easy to miss. Indeed, I missed it for many years.

To get there: From the turn-off for Cascade Lake Falls, continue driving southwest on Route 73 for another ~0.3 mile. When you reach a sign on your left that says "Town of Keene 1808," look quickly across the lake to get a glimpse of the cascade as it plunges down a high, forested rock wall. Since there is no way to pull off here because of guardrails on both sides of the road, a quick glance of the falls is all you are going to get.

The ideal way to view Cascade Lake Falls #2 in a more leisurely fashion is to take along a canoe or kayak and paddle across Upper Cascade Lake to the waterfall. Just take into account that you will not be able to escape the constant drone of cars moving along Route 73 once you are on the water.

ADIRONDACK LOJ—HEART LAKE AREA

The Adirondack Loj, overlooking 640-acre Heart Lake (originally known as Clear Lake and most likely a post-glacial kettle lake), is a fabulous hiking Mecca located in the heart of the High Peaks Region. It has been called the "finest square mile," and rightfully so. The original Loj (a Dutch word that phonetically sounds like "lodge") was built in 1890 by Henry Van Hoevenberg, who owned it until 1898. Van Hoevenberg's three-story-high lodge contained sixty rooms and was much larger than the present one. There was even a 70-foot-high observation tower. In 1900 the lodge was taken over by the Lake Placid Company, which operated it while wisely allowing Van

Hoevenberg (who had suffered a series of financial setbacks) to manage it. A terrible fire destroyed the lodge in 1903, however, and the land remained essentially unused for another two decades.

The present Adirondack Loj was built in 1927 by the Lake Placid Club. It was taken over by the Adirondack Mountain Club in 1957 and today is able to accommodate 38 guests. Reservations can be made at (518) 523-3441.

The world-renowned Adirondack Loj.

The High Peaks Information Center (HPIC) is located next to the parking area and serves as a year-round resource for hikers, backpackers, snowshoers, and cross-country skiers. Staff can provide information on weather forecasts and trail conditions. Hiking items that you forgot to include in your pack can be purchased at their store. Cross-country skis, snowshoes, microspikes, and bear canisters can be rented.

To get there: From Lake Placid (junction of Routes 73 & 86) drive southeast on Route 73 for ~3.3 miles and turn right onto Adirondack Loj Road.

From Keene (junction of Routes 73 & 9N North) drive west on Route 73 for ~10.7 miles and turn left onto Adirondack Loj Road.

Proceed south on Adirondack Loj Road for 4.8 miles and turn left into the parking area near the High Peaks Information Center (44°10.971'N 73°57.817'W). There is parking for about 200 cars. On a busy weekend this entire area for parking can be filled to the brim.

Note: Parking is not allowed on the final mile of Adirondack Loj Road—obviously an attempt to limit the number of hikers in an area that often seems to exceed its capacity.

86. KLONDIKE BROOK FALLS

A number of pretty cascades, flumes, waterslides, and chutes have formed on Klondike Brook, a medium-sized stream that rises from Klondike Notch and flows into South Meadow Brook at South Meadow. Klondike Notch, sometimes mistakenly called Railroad Notch, lies between Yard Mtn. (4,009') and Howard Mtn. (3,848').

I explored two sections of Klondike Brook and, strictly by chance, found waterfalls at both locations. It seems likely that other sections of the stream may be waterfall-bearing as well. If you wish to find out for yourself, consider this an opportunity to do a little exploring on your own without knowing exactly what you will find in the end.

This is what I found:

Section #1—The most distinctive feature on this section of Klondike Brook is an enormous, rounded, 10-foot-high mound of streambed about 100 feet long and 50 feet wide. It easily shows up on Google Earth. The stream arcs around it, following a long waterslide that drops 30 feet.

Immediately afterward, Klondike Brook races down a 100-foot-long flume chute, dropping another 30 feet in the process. It is a very dynamic and thrilling spectacle as the stream thunders past like a freight train.

Section #2—This section consists of two sites separated by a couple of hundred feet. The lower site starts off with a 10-foot-high cascade that drops into a 150-foot-long chute. Here the stream is compressed against the side wall of the northeast bank as it races madly downhill. At the bottom of the chute, the stream is momentarily subdued by calm waters in a long, rectangular-shaped pool until it is again accelerated over a 6-foot-high cascade.

The upper site starts off with a 100-foot-long waterslide that drops about 30 feet over that distance. Near the top of the waterslide, the stream momentarily spreads out into three channels before consolidating in a tiny pool halfway down. From the bottom of the waterslide, Klondike Brook is then forced through a short, boulder-choked gorge where several small cascades are produced, ending with a 6-foot-high cascade.

Klondike Trail Waterslide—This fairly broad waterslide is found in two sections, the upper of which can only be seen if you climb uphill for a short distance and look around the bend. The lower section is 40–50 feet high, covering a distance of 80–90 feet. The upper section drops 60–70 feet over a distance of 100 feet. The waterslide is very pretty, and very seasonal.

Lower waterslide on Klondike Brook.

To get there: Heading south on Adirondack Loj Road, at 3.7 miles turn left onto South Meadow Road (a seasonal backcountry road) 1.0 mile before reaching the High Peaks Information Center and parking area. Drive east on South Meadow Road for 0.8 mile. At a fork where a sign says "Klondike Lean-to 2.75 miles," bear left and continue east for another 0.2 mile to the parking area for Klondike Notch (44°11.482'N 73°56.168'W).

Note: If you are undertaking this hike in the winter, be aware that South Meadow Road is closed to traffic and becomes a ski trail. You will then have to add another 1.0 mile to your hike each way.

Proceeding on foot, cross over South Meadow Brook via a well-constructed footbridge and head southeast, following the red-blazed Klondike Notch Trail. At 0.4 mile you will pass by the Mr. Van Ski Trail (named after Henry Van Hoevenberg) coming in on your left. Continue straight ahead on the Klondike Notch Trail.

Upper waterslide on Klondike Brook.

Section #1—Somewhere around 1.0 mile from the start, you will come to the second of two planked ski bridges that span tiny creeks (44°11.103'N 73°55.360'W). From the second ski bridge follow the creek downhill, bushwhacking 0.1 mile to reach Klondike Brook, where you come out just below the main fall (44°11.149'N 73°55.290'W).

The mounded section of bedrock, very visible using Google Earth, is at 44°11.105'N 73°55.280'W. The flume chute is at 44°11.149'N 73°55.290'W.

Section #2—To reach the second section of waterfalls, you can either bushwhack back up to the trail and descend at 44°10.961'N 73°55.217'W (which is nearly straight across from a prominent hill on the opposite side of the river) or bushwhack upriver from Section #1 to Section #2, a distance of over 0.2 mile. Either way you will reach the lower set of falls at 44°10.974'N 73°55.089'W. The upper set of falls start at 44°10.928'N 73°55.029'W.

Klondike Trail Waterslide—At roughly 2.1 miles from the trailhead (or ~0.4 mile before the Klondike Dam Camp Lean-to is reached), the trail crosses a large waterslide (44°10.364'N 73°54.748'W) on a small stream emanating from the west shoulder of Howard Mtn. (3,848'). The upper section of the waterslide is not visible from the trail and can only be seen if you follow the creek uphill for a hundred feet.

For those who are not interested in bushwhacking down to the falls on Klondike Brook, the waterslide on this small brook is the only cascade that can be seen while walking along the Klondike Brook Trail.

Klondike Dam Camp Lean-to—At 2.5 miles the trail arrives at the Klondike Dam Camp Lean-to (44°10.210'N 73°54.276'W) near the confluence of two small streams. There are rapids here and lots of sounds of whitewater but, alas, no cascades.

87. ROCKY FALLS

Rocky Falls is a 6–8-foot-high flume fall formed on Indian Pass Brook, a medium-sized stream that rises from Scott Pond and flows into the West Branch of the Ausable River north of Heart Lake. The waterfall is contained in a miniature chasm surrounded by much exposed bedrock. Its name comes from the rubble of rock surrounding the swimming hole across from the base of the fall. Like a number of other Adirondack waterfalls, including Bushnell, Indian, and Rainbow, Rocky Falls is formed in a dike where less resistant rock was eroded out by the action of the stream.

A photograph of the waterfall can be seen in Den Linnehan's *Adirondack Splendor* (2004) and in *Heaven Up-h'isted-ness! The History of the Adirondack Forty-Sixers and the High Peaks of the Adirondacks* (2011), taken by Barbara Harris.

During the heydays of Adirondack lumbering, a logging camp named Hennessy's operated upriver toward Indian Pass.

At one time Rocky Falls was the starting point for hikers summiting Street Mtn. (4,166') until the trail was rerouted to its present location.

The Indian Pass Trail leading toward Rocky Falls was used by Henry Van Hoevenberg in 1903 to escape from a fire that had consumed the area. Van Hoevenberg barely made it out alive, having delayed his departure from the Adirondack Loj until the very last moment.

Rocky Falls drops through a small flume into a pool surrounded by large rocks. Photograph by John Haywood.

To get there: From the High Peaks Information Center, follow the red-blazed Rocky Falls/Indian Pass Trail as it goes around the north end of Heart Lake and then heads southwest. At 2.1 miles you will see a spur path to your right that takes you quickly west to the Rocky Falls Lean-to and waterfall (44°10.319′N 73°59.926′W).

88. MACINTYRE FALLS

MacIntyre Falls, aka MacIntyre Brook Falls, is a mottled-looking, 30–35-foot-high, 2-tiered, seasonal cascade formed on MacIntyre Brook, a small stream that rises from the north shoulder of Wright Peak (4,580′) and flows into the West Branch of the Ausable River. In *Exploring the Adirondack 46 High Peaks* (1996), James R. Burnside writes, "Just beyond, the curve of a protruding rock wall amplifies the sound of a stair step waterfall coming into view."

MacIntyre Falls is rarely missed by hikers as they make their way up to Algonquin Peak and Wright Mountain. Photograph by Amy & Chris Bergman.

In *Geology of the Adirondack High Peaks* (1986), Elizabeth Jaffe and Howard Jaffe write: "At 3255' (993 m) you will reach a waterfall, probably the last water on the trail where the county rock is very granulated anorthosite. Waterfalls often imply dikes, and this one is no exception: several 12" to 18" (30–45 cm) diabase dikes (made of fine-grained, dark igneous rock) crosscut the anorthosite and their weathering-out is the cause of the waterfall."

The brook and nearby mountain range are named for Archibald MacIntyre (McIntyre was how *he* spelled it), a state comptroller who cofounded the Elba Iron & Steel Manufacturing Company at North Elba and the iron mining operation at Tahawus in the 1830s.

A photograph taken of the waterfall by James Swedberg can be seen in the 13th edition of ADK's *Adirondack Trails: High Peaks Region* (2004). A winter shot of the fall encased in ice is shown in Cliff Reiter's *Witness the Forever Wild: A Guide to Favorite Hikes around the Adirondack High Peaks* (2008).

Because of MacIntyre Falls' trailside location and the popularity of Algonquin Peak (5,115') and Wright Peak (4,580') as hiking destinations, this is probably one of the more frequently seen waterfalls in the Adirondacks.

I'm told that at the back of the MacIntyre Falls campsite is a plaque erected in 1969 honoring the deceased pilots of the B-47 bomber that crashed on Wright Peak in 1962.

To get there: From the High Peaks Information Center, follow the blue-blazed Van Hoevenberg Trail south for 1.0 mile. When you come to a junction, continue straight ahead on the yellow-blazed Algonquin Trail and proceed southwest for 1.5 miles, or a total of ~2.5 miles. MacIntyre Falls is to your left, just upstream on a small creek crossed by the trail (44°09.566'N 73°58.776'W). A designated campsite lies nearby.

89. INDIAN FALLS

Indian Falls is a 25-foot-high waterfall formed on Marcy Brook, a medium-sized stream that rises from the northwest shoulder of Little Marcy (~4,727') and flows into Marcy Brook. The waterfall is surrounded by peaks over 3,800 feet tall and is itself located at an elevation of ~3,600 feet, making it even higher up than Fairy Ladder Falls.

In *Exploring the Adirondack 46 High Peaks* (1996), James R. Burnside writes, "Here, Marcy Brook sweeps down through a widening clearing and, at the brink of a small precipice, cascades over the chiseled lip of a huge, rounded, rock ledge."

The waterfall has always been a convenient rest stop for hikers making their way up to the summit of Mount Marcy, as well as affording spectacular views of the MacIntyre Range, which includes Algonquin Peak in the center, Wright Peak to the right, and Iroquois Peak to the left.

A photograph of the falls, taken from the base, can be seen in George Wuerthner's 1988 book, *The Adirondacks Forever Wild*.

The word "Indian" as part of a place name has been used many times in the Adirondacks—Indian Pass (aka Adirondack Pass), Indian Lake, Indian Brook, Indian Pass Brook, Indian Creek, Indian Point, Indian Mountain, Indian Head, Joe Indian Pond, Indian River, and so on. There is a strong association between the words "Indian" and "wilderness." Indian Fall's name, however, strikes me as somewhat ironic as there is no concrete evidence that Native Americans ever spent much time in the higher reaches of the Adirondack Mountains. On the other hand, what better name could there be for a waterfall directly across from Algonquin Peak (a summit given an Indian name by Verplanck Colvin in 1873).

The historic, refurbished Marcy Dam and footbridge—now gone.

In *Mount Marcy: The High Peak of New York*, Sandra Weber writes that Indian Falls was first named Wallace Falls by Bill Nye after guiding Edwin R. Wallace (an early Adirondack author) past the falls up to the summit of Mount Marcy. Later, the waterfall was called Crystal Falls (a name suggested by Henry Van Hoevenberg) for a brief period of time.

During the late nineteenth century the J. & J. Rogers Company of Ausable Forks maintained a lumber camp near the fall. A rudimentary road came up to it from the valley. The camp served as an interior outpost. During the spring, logs could be sent downstream by controlled volleys of released water.

It is likely that Indian Falls served as the base camp for the first recorded ski ascent of Mount Marcy in 1914 by Irving Langmuir (a Nobel Prize–winning chemist at General Electric) and John S. Apperson (an environmental activist who also worked at General Electric).

Up until recently, Marcy Dam was a favorite spot along the hike up to Indian Falls and Mount Marcy. The initial dam, which was part of a logging camp that closed in 1922, was rebuilt in the 1930s by the Civilian Conservation Corps (CCC). The dam impounded Marcy Brook to create a strikingly beautiful pond that became a favorite camping area and base camp for hikers. Unfortunately, after being refurbished in the early 1970s, the dam was demolished by Tropical Storm Irene in 2011, causing all the stored water to drain out and returning Marcy Pond to a marshland. For better or worse, the State Department of Environmental Conservation has decided not to rebuild Marcy Dam, believing that the concept of "forever wild" (which Marcy Dam noticeably violated) should predominate. Today the hiking trail that once crossed over Marcy Dam has been rerouted downstream.

Lateral view of Indian Falls.

I remember one incident at Marcy Dam only too well. We were leading a group of waterfall enthusiasts up to Indian Falls and stopped at Marcy Dam to take a breather. One of our members, who was videotaping the trip, set up his camera on a tripod at the dam to capture the image of Mount Colden reflected in the still waters of Marcy Pond. It was the perfect spot for a trophy photograph, but his $600 camera fell off the tripod, bounced, and went right over the side of the dam into Marcy Brook. It was unrecoverable. All I could think to say, half in jest, was that a century from now someone will find the camera downstream in Marcy Brook and by then it will be an antique.

To get there: From the High Peaks Information Center, follow the blue-blazed Van Hoevenberg Trail south for 2.3 miles to the former Marcy Dam site. From here continue southeast on the Van Hoevenberg Trail for another 2.1 miles (a total of 4.4 miles from the start) to Indian Falls (44°08.425′N 73°55.735′W), which is reached as soon as you cross over a footbridge spanning Marcy Brook. This trail, which ultimately leads to the summit of Mt. Marcy, is one that Henry Van Hoevenberg cut in 1880. Be prepared for a substantial climb from Marcy Dam to reach the waterfall.

From the top of Indian Falls, the view of the MacIntyre Range is unbeatable.

90. AVALANCHE PASS FALLS

NO

There is (or was) a most unusual and unique waterfall in Avalanche Pass that, for lack of a better name, I am calling Avalanche Pass Falls. I first read about the waterfall in the Spring 1972 issue of *Adirondack Life*. The article included a photograph of the fall taken by Clyde Smith. Ever since then I have wondered if the waterfall still exists. You see, occasionally there are rock-slides in Avalanche Pass (as the name would suggest), and one of them in 1999 (27 years after Smith's article) raised the height of the pass by 25–30 feet and may have wiped out the waterfall—or maybe not.

The waterfall has been repeatedly photographed. The same photograph of Avalanche Pass Falls can be seen in Clyde Smith's 1976 book, *The Adirondacks*. Later, a photograph of what may be Avalanche Pass Falls appeared in Lincoln Barnett's *The Ancient Adirondacks: The American Wilderness*, but taken from a greater distance back with parts of the waterfall now visible much higher up the mountain. The caption mentions that the waterfall is on the Avalanche Pass side of Mount Colden. Most recently, a picture of the waterfall appeared in James R. Burnside's 1996 book, *Exploring the 46 Adirondack High Peaks*, that predates the 1999 avalanche.

Avalanche Lake.

But why am I so interested in this particular waterfall?

Imagine encountering a small cascade on the northwest side of Mount Colden on a height of land in a pass between Mount Colden and the MacIntyre Range that is seasonal, drawing upon a very limited watershed. But come spring or following episodes of heavy rainfall, this waterfall momentarily turns into a torrent. It is then that an amazing thing happens. A large rock near the top of the waterfall cleaves the stream into two rivulets. One rivulet races southward down to Avalanche Lake, then into Lake Colden, and from there across the Flowed Lands, becoming part of the Opalescent River—a stream that flows into the Hudson River and is ultimately discharged into the Atlantic Ocean next to Coney Island.

The other rivulet, however, takes off northward, flowing into Marcy Brook, then into the Ausable River which, in turn, flows into Lake Champlain and from there into the St. Lawrence Seaway and, ultimately, into the Atlantic Ocean hundreds of miles northeast of Maine.

Think about it—a waterfall whose divided waters end up in two different sections of the Atlantic Ocean, separated by hundreds of miles. Imagine standing below, watching a piece of bark being swept over the falls, and realizing that there is no way to know at that one moment in what part of the ocean it will end up. What a perfect example of chaos theory at work or, as Clyde Smith writes, "I daresay there are few places on earth where nature has arranged a waterfall's division with such precision."

This waterfall is so unique that it was also pointed out in Paul Jamieson & Donald Morris's *Adirondack Canoe Waters: North Flow* (1986).

Avalanche Pass, containing Avalanche Pass Falls, is described in *Guide to Adirondack Trails: High Peaks Region*, edited by Tony Goodwin, as "probably the most spectacular route in the Adirondacks."

To get there: From the High Peaks Information Center, follow the blue-blazed Van Hoevenberg Trail south for 2.3 miles to the former Marcy Dam site. Take the yellow-blazed Avalanche Pass Trail that leads southwest to the top of Avalanche Pass after another 1.6 miles (or a total of 3.9 miles from the start of the hike).

I have not been back to look for the waterfall, but those hiking through Avalanche Pass might want to take a few moments to see if a waterfall still exists with the unusual properties described.

91. TRAP DIKE FALLS

Nb

There is a huge cleft in the northwest shoulder of Mt. Colden (4,714') called the Trap Dike, which Jerome Wyckoff describes in his 1967 booklet, *The Adirondack Landscape*, as "The most famous of all Adirondack dikes ... the garnet-bearing mass of metagabbro marked by the great cleft in Mount Colden." Elizabeth Jaffe and Howard Jaffe, in *Geology of the Adirondack High Peaks* (1986), write, "On the opposite shore of the lake here is a large slot in the cliff, with a tree-covered delta of tumbled-down rocks at its base."

The dike was first observed by Ebenezer Emmons in 1836. Although no marked trail leads up through this enormous vertical slot that, to my eyes, resembles a rock-filled chasm standing on its end, people do consistently use the Trap Dike to ascend to the top of Mount Colden, emulating Robert Clarke and Alexander Ralph who made the first recorded ascent in 1850. (Mount Colden was named after David C. Colden, who was involved with the McIntyre iron works.)

The Trap Dike as it looked in the 1980s before trees and brush were swept away.

In his book *Uneven Ground* (1992), Paul Jamieson describes the dike as "a cleft running up the west face of Avalanche Lake. Wide and deep at the bottom, it tapers in both dimensions toward the top." Jamieson then goes on to describe a hike that he and several companions undertook when water was raging down the dike as though it were a flume.

Another detailed account of a hike up through the Trap Dike, written by Gloria Daly, can be found in *Adirondack Peak Experiences: Mountaineering Adventures, Misadventures, and the Pursuit of "The 46"* (2009). The story includes a photo of the lower section of the Trap Dike by Neil Luckhurst.

Whether the trek up through the Trap Dike constitutes a hike or an actual climb continues to be a matter of some debate. Either way, it can be a dangerous, even potentially deadly, climb. In his 2008 book, *At the Mercy of the Mountains: True Stories of Survival and Tragedy in New York's Adirondacks*, Peter Bronski narrates the incident of a young woman who fell to her death while climbing with four friends. This is not an isolated incident. In 2011, a 22-year-old man also died while climbing up the dike.

The reason why I bring the Trap Dike to your attention is that a stream runs down through it, producing a series of cascades. These can all be seen from the west side of the lake, particularly early in the spring when an increased volume of water is flowing down from the mountain. One of the cascades is described as being 30 feet in height.

The Trap Dike as it looks today. Photograph by Amy & Chris Bergman.

To get there: If you have already made the effort to get to Avalanche Pass, why not continue then on the yellow-blazed trail until you reach the west side of Avalanche Lake, a body of water named by William C. Redfield (a meteorologist who participated in the first recorded ascent of Mount Marcy and for whom Mount Redfield is named)? The hike along Avalanche Lake is an adventurous one, where ladders and bridges take you across open spaces and around boulders.

From Marcy Dam (see "Avalanche Pass Falls") follow the yellow-blazed trail southwest for 2.5 miles until you are three-quarters of the way along the west side of Avalanche Lake. Here you will come to the second of two planked boardwalks bolted to the cliff face—helpful walkways known as "Hitch-up Matilda" (44°07.952'N 73°58.127'W). They provide a way of continuing south without having to wade through waist-deep water, as a nineteenth-century woman named Matilda fretted about doing in the 1860s.

Look directly across the lake for spectacular views of the Trap Dike as it rises up hundreds and hundreds of feet from the floor of the valley.

For those who wish to get closer to the dike, follow the yellow-blazed trail to the south end of the lake and then bushwhack north on a herd path along the shoreline until you reach the base of the Trap Dike where huge jumbles of rock rubble have collected (44°07.903′N 73°58.053′W). Although the overall view of the Trap Dike is not greatly improved, you may get a greater sense of intimacy with this geological phenomenon that contains one of the most unique waterfalls in the Adirondacks.

92. FALLS ON TRIBUTARY TO LAKE COLDEN

No

A series of falls have formed on an unnamed tributary to Lake Colden, a small stream that rises from the east shoulder of Boundary Peak (4,829′) and descends steeply into the valley below. The cascades begin virtually from above the trailhead as you hike northwest up the mountain. They include small-to-medium-sized cascades, waterslides, a plunge fall, and one particularly nice flume falls where the stream drops 25–30 feet. As Barbara McMartin writes in *Discover the Adirondack High Peaks* (1989), "Within 100 feet of elevation gain, a wonderful series of waterfalls begin. Paths lead to the pools below each chute." In her book, McMartin expresses the belief that the stream deserves a name ("it is far too pretty to remain nameless") and suggests calling it Algonquin Brook. Personally, I would favor Boundary Brook, since the stream seems to be emanating more from that area than from Algonquin Peak.

To get there: From Hitch-up Matilda (see "Trap Dike Falls") proceed southwest along the yellow-blazed trail to the end of Avalanche Lake, a distance of 2.6 miles from Marcy Dam. Cross over the outlet stream and continue following the trail southwest. At 2.9 miles bear right, now following a blue-blazed trail that leads to and along the northwest side of Lake Colden.

At 0.4 mile from the junction, you will cross over a small footbridge and come immediately to the 2.1-mile-long, yellow-blazed Boundary Trail that heads steeply and relentlessly up the mountain to the summit of Algonquin

Peak, an ascent of 2,300 feet. The trailhead is only 0.2 mile northeast of the Lake Colden DEC Interior Outpost (44°07.452′N 73°58.824′W).

The Flume Falls lies next to the Boundary Peak Trail. Photograph by Amy & Chris Bergman.

Take heart, for you will only need to hike the first 1.0 mile of the Boundary Trail to see all of the waterfalls. Still, it involves an ascent of nearly 900 feet.

Note: For waterfall enthusiasts who still have not had enough excitement for the day, a red-blazed trail at the south end of Lake Colden near the Colden Dam (44°07.163′N 73°58.958′W) leads southeast steadily uphill, paralleling the Opalescent River, where a battery of medium-sized waterfalls can be seen once you climb higher. These pretty waterfalls are not as well known as their counterparts, 15-foot-high Opalescent Falls and 75-foot-high Hanging Spear Falls, which are found on a section of the Opalescent River downstream from the Flowed Lands.

One of many waterfalls on the upper Opalescent River.

IX: NORTHVILLE-PLACID TRAIL

Wanika Falls in the fall.

The 133-mile-long, blue-blazed Northville-Placid Trail was begun by the newly formed Adirondack Mountain Club (ADK) in 1922 and completed in 1924, guiding hikers through one of the wildest and most remote parts of the Adirondacks. At its inception the trail was called the Long Trail because of its length, but that name was soon abandoned for the more descriptive name, Northville-Placid Trail, which defined its two endpoints.

In 1927 the Northville-Placid Trail (sometimes abbreviated NTP, NPT, and N-P Trail) was donated to New York State, whose Department of

Environmental Conservation (DEC) took over responsibility for its maintenance. Earlier, the north end of the Northville-Placid Trail began farther south at the end of Averyville Road (named for Simeon Shipman Avery), at which time the trek to Wanika Falls was only around 3.5 miles long. However, limited parking space as well as other issues compelled the DEC to relocate the trailhead in 1977 to its present location next to the Averyville Bridge spanning the Chubb River.

In 2014 the trailhead at the south end of the N-P Trail was relocated from Benson to Riverfront Park in Northville and, at that very moment, the footpath really did become the Northville-Placid Trail.

This chapter on the Northville-Placid Trail describes several interesting waterfalls that have formed on the Chubb River, a robust stream that rises from the north shoulder of Street Mtn. (~4,140') and flows northeast into the West Branch of the Ausable River east of the Lake Placid Airport. The river, as well as Chubb Hill on Old Military Road, was named after a North Elba settler named Joseph Chubb who owned a large tract of land off of Old Military Road. Unlike other tributaries to the Ausable River, the Chubb River has a darker, tannin look to it because of the proliferation of lush vegetation along its floodplains.

The lower part of the Chubb River was industrialized as early as 1809 by Archibald McIntyre, who erected the Elba Iron Works. As far as I know, the upper parts of the river were left untouched. If you drive along Route 73 today and look to your left just after Station Street, 0.2 mile before reaching Route 86, you will see a dam on the Chubb River called Jack's Dam (named after John D. "Jack" Barry), a relic from bygone industrial days (44°16.641'N 73°58.999'W). I've read that the Chubb River is the only major tributary to the West Branch of the Ausable River, all the others being considered minor streams.

To get there: From Keene (junction of Routes 73 & 9N North) take Route 73 west for ~12.4 miles. Turn left onto Old Military Road and head northwest for 1.6 miles. When you come to Averyville Road, turn left and proceed southwest for over 1.1 miles. Just before crossing over the Chubb River, turn left into the parking area for the Northville-Placid Trail (44°15.771'N 74°00.817'W).

From Lake Placid (junction of Route 73 & 86) drive southwest on Route 73 for 0.2 mile. Then turn right onto Station Road and proceed southwest for 0.8 mile. When you come to Old Military Road, turn right and then immediately left onto Averyville Road. After 1.1 miles, turn left into the parking area for the Northville-Placid Trail.

93. CHUBB RIVER CHASM

The Chubb River Chasm is a small flume that has formed on the Chubb River. The flume is a unique feature of the river that demands not only recognition, but its own identity. For this reason I have named it Chubb River Chasm. The chasm is approximately 30 feet long and partially filled with boulders.

Directly under the footbridge is an 8-foot-high waterfall that consists of several small drops, culminating in a final 4-foot-high cascade that launches itself into a shallow pool of water. Two feet from the base of the fall is an oblong slab of rock that rises up from the pool like some kind of threatening monolith.

Small cascades tantalize at the Chubb River Chasm.

To get there: From the parking area, head north on the very well maintained, blue-blazed Northville-Placid Trail. The Chubb River quickly disappears from sight and won't be seen again until you reach Chubb River Chasm. In between is a hike of approximately 6.0 miles that can grow wearisome after several miles. Hang in there, for the end result is well worth the energy expended. Although you will cross over a number of tributaries to the Chubb River, none are substantial in size and none are waterfall-bearing within eyeshot of the trail. To be sure, some of these tributaries do contain cascades, but you would have to bushwhack off-trail and be willing to add on extra miles to your hike in order to check out the possibilities. In Dean S. Stansfield's 2003 book, *Adirondack Venture: Images of America*, a photo taken in 1905 of four hikers shows them crossing a cascade on one of these tributaries.

Approaching the 6.0 mile mark you will pass by a trailside campsite and then, within another 100 feet, reach a railed footbridge that crosses over the Chubb River directly above the chasm (44°12.458'N 74°03.383'W).

When you cross over the footbridge, take note that the old Northville-Placid Trail used to come in at the end of the bridge.

94. MINOR CASCADES ON THE CHUBB RIVER

NO

Between Chubb River Chasm and Chubb River Falls are a number of minor cascades that can be seen from the trail. I suspect that many of these simply vanish in the early spring when the river is engorged and becomes one big blur of water.

The one waterfall that caught my attention consists of a series of small cascades in fairly close proximity to one another, dropping a total of 10 feet over a distance of 50 feet.

To get there: From Chubb River Chasm, cross over the footbridge and continue following the trail as it now parallels the south side of the river. Within 0.05 mile you will see to your left a series of tiny cascades (44°12.316′N 74°03.279′W). Although no path leads down to them, a quick scramble will get you to the streambed for a closer look if you so desire.

95. CHUBB RIVER FALLS

NO

Chubb River Falls consist of two large cascades separated by a distance of 70 feet. When you stand at the base of the lower cascade and look up, however, the two seem to merge together as one large, two-tiered waterfall.

The lower fall is 25 feet high and made of a very rough-cut gleaming rock. It is much steeper than the upper fall. The upper fall is 30–40 feet high and formed in two drops. Between the two main falls is a section of highly tilted bedrock.

It seems reasonable to call these two large cascades Upper Chubb River Falls and Lower Chubb River Falls so that they can be differentiated from Wanika Falls, which lies farther upstream. Undoubtedly many hikers have stopped at Chubb River Falls thinking that they have reached Wanika Falls—despite the sign at the junction that gives the mileage to Wanika Falls as 0.2 mile. Even if these hikers then followed the spur path past the first two falls, they would quickly reach the point where the path comes up to the Chubb

River and, seeing nothing there, assume that they had just passed Wanika Falls.

Thirty-five-foot-high Upper Chubb River Falls.

By giving a name to the first two large waterfalls on the Chubb River, I hope that hikers will not turn back prematurely before reaching Wanika Falls.

To get there: From the Minor Cascades, continue following the blue-blazed Northville-Placid Trail as it leads gradually uphill, paralleling the Chubb River. After 0.5 mile from the chasm footbridge, a junction is reached. A sign points out that you have now hiked 6.5 miles from the trailhead parking. Another sign points the way left to Wanika Falls on a very well-worn spur path.

Follow the spur path to reach Lower Chubb River Falls within 150 feet. There is a faint, informal path leading down to the base of the waterfall that can be taken, but the informal path is really more of a bushwhack than an actual path. At the base of lower Chubb River Falls, the GPS reading is 44°11.997'N 74°03.448'W.

There is no faint path leading uphill to the upper fall, but you can bushwhack to it from the base of the lower fall if you don't mind scrambling along the sloping side of the gorge. The GPS reading at the base of the upper fall is 44°11.958'N 74°03.431'W.

If conditions are right for crossing over the Chubb River near the top of the falls, there may be better, more established paths on the opposite side that lead to views of Chubb River Falls.

96. WANIKA FALLS

NO

Wanika Falls is a huge waterfall, easily 150 feet in height. Pictures of the waterfall simply do not do it justice.

The waterfall contains several large sections with significant drops. I saw the falls when only a modest flow of water was cascading down. It's hard to imagine what it must look like in mid-spring or early summer. As I stood by the pool at the base of Wanika Falls and looked up, the top of the waterfall almost seemed to vanish into the clouds.

The first thing I noticed about Wanika Falls was the enormous volume of space it occupied. Wanika Falls is not only a tall waterfall, it is massive and wide. Nothing that you have already seen along the trail will prepare you for the sight ahead as you step out of the confines of the forest.

Over the years Wanika Falls has become a favorite camping site for hikers trekking along the Northville-Placid Trail. For hikers who are not doing large sections of, or the entirety of, the Northville-Placid Trail, a round-trip day hike of nearly 14.0 miles makes for few takers. Were it not for the Northville-Placid Trail, I suspect that Wanika Falls would remain a fairly obscure, rarely visited waterfall in an area far removed from any roads. It would essentially be unvisited, and that would be a terrible shame. On the other hand, if Wanika Falls was only a tenth of a mile from a main highway, it would certainly be one of the premier, most heavily visited attractions in the Adirondacks, and that would also be a shame for it would become overused. Perhaps the right balance has been struck here—a magnificent waterfall that is accessible if you want to make the effort to get to it.

The word Wanika apparently is of Hawaiian origin, meaning "God's gracious gift." It's not entirely clear to me how a Hawaiian name came to be applied to an Adirondack waterfall.

Wanika Falls flows down like water pouring out of the sky.

I have seen few photographs of Wanika Falls, which suggests that not many photographers have been to it. The photograph of Wanika Falls that I am most familiar with was taken by Richard Nowicki and included in the 4th Edition of the *Northville-Placid Trail* guidebook.

In Den Linnehan's 2004 photo book, *Adirondack Splendor*, several pages of photographs are devoted to Wanika Falls, but I believe they are actually of Chubb River Falls—a good thing, I believe, for Chubb River Falls is deserving of recognition, too.

To get there: From Chubb River Falls continue following the spur path for another couple of hundred feet. You will come to a point where the trail seemingly ends near the top of Chubb River Falls. Look across the river and you will see that the trail actually continues on the other side. If you are visiting later in the season (which I was), you can easily cross over to the opposite side. It's really just a matter of stepping on a few rocks across the river and then onto dry bedrock. If you are visiting in the early spring, however, you really need to stop and give some thought as to just how you are going to cross the river, or even if you should. Let's take a look at the harsh realities. The trail crosses the river virtually at the top of Chubb River Falls, where there is smooth, inclined bedrock. Should the river sweep you off your feet, you would be carried right over the top of the falls with nothing to grab onto to arrest your slide.

If you must cross when a fair amount of water is flowing, I would recommend making your way upstream for 50–75 feet and then crossing where the streambed is rockier. At least then you would have a chance to drag yourself over to the bank before being swept over Chubb River Falls if you lost your footing. I would highly recommend testing the water first to see how strong the current feels before you commit fully to the crossing. I would also recommend using water shoes and hiking poles for better footing and balance (and to spare your hiking boots from getting soaked, since you will need them for the hike back). With all this said, I simply wouldn't attempt a high-water crossing unless I was upstream a safe distance from the top of the falls and felt confident that I could manage the current. Better to just wait until summer and visit when the stream has quieted down. That way you avoid all the hassles and dangers involved in doing an early-season crossing.

But let's assume that you have crossed over the Chubb River and are now at the streamside campsite. At one time there used to be a shelter here, but that was some time ago. Take note of paths going off to your left that take you downstream.

Just upstream from the campsite, less than 50 feet away, is a pretty, 6–8-foot-high cascade—an extra bonus, I suppose, if you are camping at the site.

To get to Wanika Falls from the campsite, follow the well-worn trail to your right that leads upstream. In less than 0.1 mile you will emerge from the woods at the base of Wanika Falls (44°11.912'N 74°03.353'W). Be prepared for a mind-blowing experience. Having just seen a substantial waterfall downstream at Chubb River Falls, it is hard to imagine that anything even larger could possibly exist farther upstream.

As I hiked out from Wanika Falls back to my car, I was left with the feeling that I had seen something special. That night, after a hike of nearly 14.0 miles, it was "boots and roots," and "rocks and socks" as I tried to nod off to sleep. I don't know if you've had this strange experience (like a song that's wormed its way into your mind), but after looking at the trail all day long and seeing only my boots as I scanned the terrain to avoid rocks and roots, those very same images kept dancing across my eyes as I lay in bed. A small price to pay for such a wonderful experience, for sure.

X: WILMINGTON NOTCH

Wilmington Notch is a formidable mountain pass created by glaciers and, later, the West Branch of the Ausable River.

Wilmington Notch is a massive mountainous pass that was carved out of 1.5-billion-year-old rock by glaciers over 10,000 years ago. The glaciers accomplished this feat by pushing fractured rock out of a preexisting fault line, enlarging it considerably, which then allowed the opportunistic West Branch of the Ausable River to deepen and expand it even further. As a result the Wilmington Notch's northwest wall rises up to as high as 1,700 feet, and the southeast wall to over 700 feet. Racing through this passageway, the West Branch is squeezed by the mountains, causing the river to accelerate and increase in power dramatically. The notch essentially begins around five miles northeast of Lake Placid.

The West Branch rises from Mount Marcy and the MacIntyre Range, with many mountain-based tributaries feeding into it, including South Meadow Brook, Marcy Brook, and the Chubb River. It has a drainage area of 236 square miles. Bradford B. Van Diver, in *Roadside Geology of New York*, describes this section of the West Branch as "a clear, bouldery mountain stream, with bedrock banks, frequent rapids, waterfalls, and quiet pools that attract many trout fishermen in season."

The word "Ausable," aka Au Sable, is French for "of the sand," a reference to the sandy delta on Lake Champlain created where the Ausable River flows into the lake at Ausable Point.

In *The Military and Civil History of the County of Essex, New York* (1869), Winslow C. Watson writes: "The river compressed within a narrow passage of a few feet in width, becomes here an impetuous torrent, foams and dashes along the base of a precipitous wall, formed by Whiteface mountain, which towers above it, in nearly a perpendicular ascent of thousands of feet, whilst on the other side it almost laves the abrupt, naked, and ragged crags, of another lofty precipice."

After about 45 miles from its highest point on Mount Marcy, the West Branch joins with the East Branch at Ausable Forks to produce the formidable Ausable River. Before combining forces, however, the West Branch has created a series of waterfalls in Wilmington Notch—Quarry Pool Falls, Monument Falls, High Falls Gorge Falls, Wilmington Notch Falls, and the Flume Falls.

Route 86, which follows along the West Branch through the Notch, was established during the 1930 renumbering of state highways in New York, largely replacing a realigned NY 3. Route 86 heads northeast from the terminus of Route 73 at the southeast corner of Lake Placid village, eventually paralleling the West Branch of the Ausable River. In Wilmington it does a right-angle turn and then heads east to Jay, where it ends at Route 9J. In the early days, passage through the Notch consisted of a narrow, muddy road that was primarily used by lumber teamsters. It was modernized in 1916 and then again in 1955.

97. WHITEFACE BROOK FALLS

Multiple cascades have formed along Whiteface Brook, a substantial stream that rises from a cirque on the west shoulder of Whiteface Mtn. (4,867') and flows into Lake Placid near Whiteface Landing.

The first significant grouping of cascades upstream from the Whiteface Brook Lean-to shelter consists (in descending order) of an 8-foot-high cascade, a 5-footer, a 25-foot-long waterslide dropping 6 feet, and ends in a 4-footer.

Whiteface Brook Falls, located 0.2 mile farther upstream, is a massive, 30-foot-high cascade containing several drops. It is immediately followed by a 4-foot cascade and then, farther downstream, a 3-footer. At the top of the cascade is a 60-foot-long waterslide. Woe to the hiker who rides down this slide and inadvertently goes over the top of the waterfall!

Thirty-foot-high Whiteface Brook Falls is the high point and highlight of the hike.

To get there: From Lake Placid (junction of Routes 86 & 73) drive northeast on Route 86 for 2.9 miles. Turn left onto Connery Pond Road (presently a dirt road) and proceed northwest for 0.6 mile to a small parking area at the drivable end of the road (44°18.522′N 73°56.185′W). If the parking area is full, park at one of the pull-offs along the way.

Follow the red-blazed trail (an abandoned logging road that was used to clean up debris from the "Big Blowdown of 1950" with its hurricane winds of 105 MPH) north for 2.5 miles. The first 0.3 mile is a bypass route, diverting hikers away from a private residence on the lake. For virtually the entire distance to Whiteface Landing, the trail is wide and well-trodden, with little change in elevation. It makes for a very easy hike. Don't be concerned about the lack of red-blazed trail markers once you leave the Connery Pond area. There is no way to get lost on this trail. You will know that you have reached the 2.5-mile mark when you come to a junction where the sign reads, "Whiteface Landing 0.1 mile. Whiteface Brook Lean-to 1.0 mile. Whiteface Mountain 3.0 miles."

Whiteface Landing is 0.05 mile ahead and located at the east side of Barrel Bay. It provides a scenic view of the northeastern end of Lake Placid, complete with a dock and campfire pit. Evidently many people start the hike up Whiteface Mountain from this point after arriving by boat.

From the trail junction, head northeast, following the red-blazed Whiteface Mountain Trail as it slowly advances uphill. After ~0.6 mile the trail comes up to Whiteface Brook, where a series of small cascades can be seen. None are greater in height than 2–3 feet (44°20.579′N 73°55.892′W). Little time need be spent here.

Although the Whiteface Mountain Trail no longer crosses back and forth across Whiteface Brook as it once did (having caused untold problems for hikers, I'm sure), you will still have to ford a sizeable tributary coming in from the east, and this can be tricky to accomplish in the early spring or during times of high water. Fortunately, this is the only stream crossing that you will have to contend with.

You will reach the Whiteface Brook Lean-to (44°20.723′N 73°55.643′W) at ~1.0 mile. There are no cascades here, but there is a commanding, 8-foot-high, 25-foot-long, oblong-shaped boulder near the lean-to as well as a perched rock not far behind the shelter. I bring this to your attention because in 2015 I coauthored a book with photographer Christy Butler called *Rockachusetts: An Explorer's Guide to Amazing Boulders of Massachusetts* after having developed an abiding interest in large boulders and rock formations. Like waterfalls, glacial boulders add diversity and interest to a landscape that otherwise can be fairly uniform in appearance.

From the shelter continue uphill, following the red-blazed trail. Disconcertingly, Whiteface Brook quickly veers away and begins to parallel the trail approximately 100–150 feet distant. It remains out of sight for the next 0.5 mile. For this reason you may have to rely upon the GPS coordinates given in order to find the next set of waterfalls. On the other hand, bear in mind that I was able to locate these waterfalls by reading the terrain and listening closely to the sound made by Whiteface Brook, so what worked for me might also work for you.

At around ~0.3 mile from the shelter, turn left into the woods and bushwhack west for several hundred feet to reach Whiteface Brook. If all goes well you will come out to a pretty series of cascades (44°20.939′N 73°55.556′W).

Returning to the trail, continue your uphill ascent. The terrain now becomes increasingly steeper. After another 0.2 mile (or ~0.5 mile from the shelter) look for a rocky bluff in the woods to your left. Head into the forest for 150 feet and, if all goes well again, you will come out to the base of 30-foot-high Whiteface Brook Falls (44°21.061′N 73°55.506′W).

Back on the trail, continue hiking uphill. Soon the trail and stream come together again and you will find yourself walking along the rim of a high gorge. When you reach a large trailside boulder at the top of the gorge, look upstream to see a small waterslide cascade below. This is the turn-around point unless you are planning to summit Whiteface Mountain or wish to continue exploring Whiteface Brook farther upstream.

Waterfall on Whiteface Brook. Postcard c. 1940.

I undertook this hike in early November. As I followed the trail up from Whiteface Landing, autumn literally turned into winter. By the time I reached the Whiteface Brook Lean-to, the trees and ground were covered with thick snow and the temperature had dropped into the low 20s. Bushwhacking increasingly became more difficult because, every time I left the trail, snow from overhanging trees would run down my neck as well as onto the rest of me. Soon all my clothes were sopping wet. Definitely, a rain jacket with a hood would have improved my comfort level considerably.

98. HOLCOMB POND CASCADE

I have added this tiny cascade (which could just as easily be called a cascading stream) because it is nearby, roadside, and pretty. It is a pleasant spot to stop and take a break in the sunshine while enjoying the sound of gurgling water. The cascade is formed on Holcomb Pond's outlet stream.

The small park next to the cascading stream is dedicated to the 10th Mountain Division of Adirondack soldiers who served their country between 1943 and 2013. Look for a memorial plaque on the boulder next to the pull-off.

To get there: From Connery Pond Road head northeast on Route 86 for over 0.2 mile and turn right onto River Road. Proceed south for 1.0 mile. The cascading stream is on your left where its waters flow under River Road into the West Branch of the Ausable River (44°17.203′N 73°55.945′W).

If you are approaching from Route 73, turn onto River Road and drive north for 2.9 miles. The cascade is on your right.

99. QUARRY POOL FALLS

Quarry Pool Falls is formed on the West Branch of the Ausable River in an area that can't be seen from Route 86. This 3–4-foot-high, dammed fall is contained at the head of a trough-like gorge where potholes of all sizes can be seen worn into the bedrock. In the April 1989 issue of *Adirondac*, Donald Morris describes the site succinctly as "a small gorge with a cataract below the spillway of a breached dam." Paul Jamieson & Donald Morris, in *Adirondack Canoe Waters: North Flow* (1987), write: "Here the stream is constricted to 20 feet in a small gorge. Below the spillway of a breached dam is a white torrent twisting between flanks of bedrock and followed by rapids."

Don't let the size of this waterfall fool you, however. It is well worth a visit.

While exploring the area, take a moment to consider the following question: does the name "quarry pool" refer to an abandoned, water-filled pit—a leftover from past days of quarrying—or does the name refer to fish (*quarry* pursued by fishermen) that swim about in *pools* along the West Branch? You be the judge.

Quarry Pool Falls is one of the West Branch of the Ausable River's hidden treasures. Photograph by John Haywood.

Upriver, flat-water paddlers launch their canoes and kayaks from a put-in along nearby River Road and typically carry their watercraft around Quarry Pool Falls—a Class II–Class III stretch of whitewater—before resuming their paddle to Monument Falls, where they take out just before the cascade. Whitewater enthusiasts, needless to say, generally push right through the small gorge. Perhaps if you're lucky you will see some action on the river while you are there.

To get there: From Lake Placid (junction of Routes 86 & 73) drive northeast on Route 86 for 3.5 miles and turn left into a pull-off at a sign that reads "Quarry Pool" (44°18.169′N 73°55.337′W).

Follow a well-worn path that takes you down to the river in 0.05 mile. Then bear left, following the path upriver for 0.05 mile. You will come to Quarry Pool Falls at the head of a small gorge just beyond a large boulder next to the streambed (44°18.235′N 73°55.438′W). The path ends at the top of the falls.

100. MONUMENT FALLS

Monument Falls—so-named because of two large stone monuments next to the parking area—is the first waterfall you will see as you drive northeast from Lake Placid. A plaque on the 1935 monument boulder reads: "This tablet commemorates the fiftieth anniversary of conservation in New York State. On May 15, 1885, Governor David B. Hill signed the law establishing the Forest Preserve. The surrounding mountains, streams and woodlands have been acquired for the free use of all the people of the state and are maintained as wild forest land for the enjoyment of future generations." The plaque on the second monument commemorates the 100th anniversary of the Forest Preserve.

Monument Falls was named for its two large, centennial stone monuments.

When the word "monument" comes to mind, the first thing that I think of is rock, and Monument Falls has plenty of that. The waterfall is not big—perhaps 6 feet high at most—but over it glides the West Branch as it passes through a tiny gorge. The waterfall is a favorite fishing site and represents the first significant drop on this section of the West Branch. Whitewater paddlers consider it to be a Class III–IV run. Bigger drops are soon to follow.

The view from the parking area is especially scenic as you look over the top of shimmering Monument Falls with Whiteface Mountain framed in the distance.

Photographs of the waterfall can be seen in Paul L. Gibaldi's 2007 book *Spirit of the Adirondacks* and Derek Doeffinger & Keith Boas's *Waterfalls of the Adirondacks and Catskills* (2000).

Monument Falls makes up for its lack of size by the sheer majesty of its scenery. Photograph by John Haywood.

To get there: From Lake Placid (junction of Routes 73 & 86) drive northeast on Route 86 for 4.2 miles (or 0.7 mile past Quarry Pool Falls) and turn left into the parking area for Monument Falls (44°18.650′N 73°54.927′W). The parking area can get a bit crowded when fishermen are out in force.

A 200-foot-long path hugging the guardrail leads to the waterfall (44°18.696′N 73°54.913′W). There is much exposed bedrock at the fall, allowing you under normal conditions to descend to the level of the streambed.

101. OWEN POND'S OUTLET CASCADE

In all fairness a disclaimer is needed here. The cascades on Owen Pond's outlet stream (a tributary to the West Branch of the Ausable River) are barely more than rapids. There are two 2-footers and, farther upstream where the stream narrows, a 3-footer.

However, the walk along the Owen Pond Trail is very short and well worth taking just for the fun of it. In doing so you will also get to see an enormous glacial boulder at 0.3 mile engulfed by the roots of a tree that is holding onto it tenaciously for life. It is like something out of the 1979 movie *Alien*. Clearly, this was the highlight of the trip for me. A photograph of the ensnarled boulder can be seen in Den Linnehan's *Adirondack Splendor* (2004). The main cascade is over 0.05 mile up the trail from the boulder.

The trail ultimately leads to 19-acre Owen Pond, which is 30 feet at its deepest but very rocky near its outlet.

A tree tenaciously grapples with an enormous boulder, much like a giant squid clutching a whale.

To get there: From Lake Placid (junction of Routes 73 & 86) drive northeast on Route 86 for 5.3 miles (or 1.1 miles past Monument Falls). Park to your right in the pull-off (44°19.566'N 73°54.643'W) and follow the blue-blazed trail east up to the cascades, a distance of less than 0.4 mile.

The Owen Pond Trail can also be approached from the Shadow Rock Pond pull-off (44°19.382'N 73°54.754'W), 0.2 mile south of the parking area for Owen Pond. The pull-off lies next to the confluence of Owen Pond's outlet stream with the West Branch.

From the pull-off, cross over the road and follow a well-worn, unmaintained path upstream along the north side of the brook. This path was part of the original trail to Owen Pond and enters the main, blue-blazed trail within 0.1 mile. Taking it may provide you with a greater sense of entering wilderness—a semi-bushwhack, if you will. It seems likely that this former trail was discontinued because of limited parking space as well as because it necessitated crossing the road to begin the trek.

102. INTERIM POINT OF INTEREST

The West Branch passes through a narrow gorge that is contained within a larger gorge—Wilmington Notch. Towering rock buttresses are visible on both sides of the highway and stream, looming precipitously above you. The higher rock buttress is known as the Wall of Jericho, a biblical reference to a Bronze-age wall that defended the ancient city of Jericho.

This is an especially scenic spot. Although there are no cascades or even any rapids of note, you will not be disappointed if you pull in to view the lofty walls of the notch and the river going through it. It's mind-boggling to think that all of this is the creation of glaciers augmented by the cutting power of the West Branch of the Ausable River.

To get there: From Lake Placid (junction of Routes 86 & 73) drive northeast on Route 86 for 6.4 miles (or 1.1 miles past the Owens Pond trailhead). Turn left into a large pull-off (44°20.246'N 73°53.806'W).

Several short paths lead down into and along the gorge.

How often do you get to be inside a gorge that itself is inside a gorge?

103. HIGH FALLS GORGE

High Falls Gorge is a 22-acre, privately owned attraction that is one of three natural geological wonders in the Adirondacks (the other two being Ausable Chasm near Keeseville and Natural Stone Bridge and Caves in Pottersville). The gorge opened around 1890, closed after World War II, and then reopened in 1961. It has been in continuous operation ever since. Like Ausable Chasm, spectacular scenic views have been opened up to the public that would otherwise be impossible to see and enjoy while timidly peering down from the rim of the gorge far above. Stairways and catwalks make accessing the gorge and falls easy and safe and create a sense of intimacy that is simply remarkable.

The geology of the gorge was described by C.R. Roseberry in his 1982 book, *From Niagara to Montauk*: "The water crashes headlong into a mishmash of cross-faults and slabs of basaltic dikes, injected in the crush zones of faulting. An intrusion of attractive pink granite, itself chopped up with the rest of the medley, throws in a syncopation of color." It all sounds very exciting—and it is.

In writing about High Falls Gorge in *Rocks and Routes of the North Country, New York* (1976), Bradford B. Van Diver eloquently demystifies the process by which waterfalls form: "Rivers extend themselves by eroding headward and when they meet a mass of harder rock, falls are formed. Since the rate of erosion under falls is greater than elsewhere, the falls may be viewed as Nature's response to an obstacle placed in its path. In other words, the falls are an attempt by Nature to restore a condition of equilibrium represented by a smooth stream gradient without

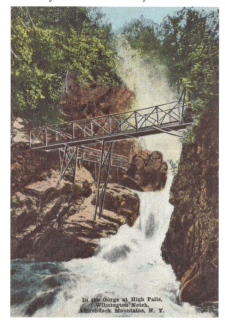

High Falls Gorge and its awesome waterfalls. Postcard c. 1940.

precipices." I find this to be an interesting way of thinking philosophically about why waterfalls exist.

There are three distinct waterfalls in High Falls Gorge. The first, the Main Fall, is 35 feet high, entering the gorge diagonally. According to Bradford Van Diver in *Roadside Geology of New York*, "The uppermost falls drops through a slot where one of these (a dark gray basaltic dike) has been worn away." The second waterfall, Rainbow Falls, is 20 feet high. The third waterfall, Climax Falls, is 25 feet high.

A glass-bottomed viewing area near the start of the gorge hike allows you to look straight down into its wild depths as the West Branch races by. It is reminiscent of the Grand Canyon Skywalk.

The gorge is reached and its interior accessed by three footbridges built in the 1960s.

High Falls Gorge is advertised as "700 feet of falls," a true enough statement, although the 700 feet actually refers to the length of the gorge, not its height. In no way does this detract from the falls being anything less than incredible.

A photograph of High Falls Gorge dating back to the late 1800s can be seen in the Wilmington Historical Society's 2013 book, *Wilmington and the Whiteface Region: Images of America*. Also included is a 1905 photo of an early version of the footbridge spanning the West Branch across from a more primitive-looking Visitor Center. More recent photographs of the gorge and falls are shown in Den Linnehan's *Adirondack Splendor* (2004), Derek Doeffinger & Keith Boas's *Waterfalls of the Adirondacks and Catskills* (2000), and Hardie Truesdale's *Adirondack High: Images of America's First Wilderness* (2005).

Make sure not to miss the 7-foot-wide, 35-foot-deep pothole along the river bank that formed when the West Branch had yet to erode deeper into the bedrock. Many other smaller potholes can also be seen throughout the gorge.

To get there: From Lake Placid (junction of Routes 73 & 86) drive northeast on Route 86 for 7.6 miles and turn left into the parking area for High Falls Gorge. Walk over to the Visitor Center (44°20.852'N 73°52.643'W) to begin the adventure.

What's especially appealing about this attraction is that your tour of the gorge is a self-guided one, allowing you to proceed at your own pace.

High Falls Gorge is privately owned, maintained, and operated by ROANKA Attractions Corp. in Wilmington. The gorge is located at 4761 Route 86, Wilmington, NY 12997. Staff can be reached at (518) 946-2278. For more information, visit their Web site at highfallsgorge.com.

104. WILMINGTON NOTCH FALLS FROM WILMINGTON NOTCH CAMPGROUND

Wilmington Notch Falls is a 40–50-foot-high cascade formed in a dynamic rocky gorge on the West Branch of the Ausable River (44°21.029′N 73°51.808′W). In the April 1989 issue of *Adirondac*, Donald Morris states, "These falls drop about 60 feet in two distinct pitches." The waterfall is featured on the cover of Derek Doeffinger and Keith Boas's 2000 photo book, *Waterfalls of the Adirondacks and Catskills*.

Wilmington Notch Falls as seen from below the Wilmington Notch Campground. Photograph by John Haywood.

Wilmington Notch Falls is impressive, but so is the overlook—a section of the gorge that projects out from the side wall toward the river. When you walk out onto this rock buttress, look across to the opposite side of the river to see a similar outthrust from the west bank. Most likely these two protrusions were connected at one time, forming a solid wall until the rock between them was worn away by the river.

What I remember most was the trepidation I felt as I ambled out onto the overlook to take a photo of Wilmington Notch Falls. Perhaps it was the combination of exposure (there's little to grab hold of) and the narrowness of this bluff. I don't know, but this was one of those times that I just couldn't shake a sense of dread. No doubt you will fare much better. I would imagine it's possible to scamper down closer to the river, but I was satisfied with the shots from the overlooking bluff and never made the attempt.

Wilmington Notch Campground has been around for a while. It opened in 1926 with just 12 campsites. Today it offers 54 campsites with all the usual amenities. It is a great place to stay if you enjoy fishing or just want to explore the West Branch.

A photograph of the waterfall, including part of the overlook, can be seen in Randi Minetor's *Hiking Waterfalls in New York* (2014).

To get there: From Lake Placid (junction of Routes 73 & 86) drive northeast on Route 86 for 8.5 miles. Turn left into the Wilmington Notch Campground, which is located at 4953 Route 86, Wilmington, NY 12997, (518) 946-7172 (44°21.054'N 73°51.651'W).

From behind the bathroom facilities, follow a path formed by run-off that leads downhill for 0.1 mile to the waterfall. It is a descent of roughly 250 feet and can be somewhat steep in places. The best viewing area is from a rocky buttress that extends out from the east bank directly downstream from the waterfall. You will be looking down from quite a height. Amazingly the falls, which are of decent height, appear even higher when seen from above, an optical illusion created by the cascade's horizontal length.

FALLS AT WHITEFACE MOUNTAIN SKI CENTER

Whiteface Mountain, blessed with a number of waterfall-bearing streams, has a fascinating history. In 1938 Herman Smith-Johannsen and members of the Lake Placid Ski Council cleared an old logging road on Little Whiteface Mtn. (3,645') and created a vertical drop of 2,700 feet for skiers. Meanwhile nearby Marble Mountain (2,753'), whose ski trails preceded Whiteface's, never proved to be financially successful, in part because it was plagued by high winds. Two seasons after Whiteface opened, Marble Mountain closed, never to reopen.

In 1955 an experimental trail approved by New York State on the eastern slope of Whiteface Mountain proved to be a smashing success. In 1958 the current Whiteface Mountain Ski Center officially opened. The ski center subsequently went through a series of expansions. Not only was the mountain's vertical increased to 3,430 feet (the greatest continuous vertical drop in eastern North America), but the ski center went through $14 million of preparations for the 1980 Winter Olympics that was hosted in Lake Placid.

Whiteface Mountain Ski Center.

The pretty bridge spanning the West Branch is called the Whiteface Access Road Bridge and was built in 1956. It overlooks some very big boulders in the stream below.

Today, the Whiteface Mountain Ski Center is more than just a ski center. It offers blazed nature trails, cloud-splitter gondola rides, a disc-golf course, an adventure zone, an airbag jump, downhill mountain biking, and 4x4 Alpine Expeditions.

To get to Whiteface Mountain Ski Center: From Lake Placid (junction of Routes 73 & 86) drive northeast on Route 86 for ~8.8 miles (or 1.2 miles past High Falls Gorge). Turn left into the Whiteface Mountain Ski Center.

From Wilmington (junction of Routes 86 & 431) drive south for ~3.0 miles and turn right into the ski center.

From either direction drive downhill, going west, for 0.1 mile and turn right into a large parking area (44°21.239'N 73°51.568'W) before crossing over the West Branch of the Ausable River.

105. WILMINGTON NOTCH FALLS

In addition to reaching the waterfall from the Wilmington Notch Campground (see "Wilmington Notch Falls from Wilmington Notch Campground"), Wilmington Notch Falls can also be accessed from Whiteface Mountain Ski Center, and in many ways this approach is the preferred one.

Wilmington Notch Falls from the west side of the river.

As mentioned previously, Wilmington Notch Falls is a massive, 40–50-foot-high cascade that has formed on the West Branch of the Ausable River (44°21.029'N 73°51.808'W). It is contained in a dynamic gorge with huge areas of exposed bedrock, massive boulders, and raging waters.

There is also a 6-foot-high flume cascade (44°21.028'N 73°52.029'W) 0.3 mile upstream from Wilmington Notch Falls that is infrequently visited.

It is hard to believe that both the falls at High Falls Gorge and the falls at the Wilmington Notch Campground share the same river with the subdued cascades at Quarry Pool Falls and Monument Falls. The West Branch is truly a river of surprises.

To reach the falls: From the parking area, cross over the West Branch via the Whiteface Access Road Bridge and immediately turn left onto a path that heads upriver behind the ski center. Go south for 0.2 mile, walking between the river and a ski tow. When you come to the end of the ski tow, bear left and proceed over to a shed next to a large, 8-foot-high glacial boulder. Between the two is the start of the red-blazed West Branch Nature Trail, a loop of 1.9 miles. The trail quickly leads down to the West Branch, where the waterfall is located. Once near the river, follow a secondary path left that takes you to a high buttress overlooking the falls. The view is somewhat limited by intervening trees. You can also scramble down to the base of the falls before walking out onto the buttress, but be careful if you do.

Back on top of the river bank, follow a secondary trail that leads upriver to the top of Wilmington Notch Falls. Be sure to take note of a dry side channel on the west side of the fall. In times of heavy water flow, part of the West Branch is diverted through this section, producing additional cascades.

106. STAG BROOK FALLS & UPPER CASCADES

Stag Brook Falls is a 40-foot-high waterfall formed on Stag Brook, a small stream that rises from the east shoulder of Whiteface Mtn. (4,867'), New York's fifth highest mountain, and flows into the West Branch of the Ausable River.

There are at least fifteen distinct cascades upstream from Stag Brook Falls, making this stream a stairway to heaven for waterfall enthusiasts. Although Stag Brook is a small stream, it seems to be powered by a good-sized watershed and never runs dry. Even in the early fall after months of little rain, the brook was still flowing and reveling in its cascades as I followed it upstream.

Stag Brook Falls is one of fifteen waterfalls formed on Stag Brook. Photograph by John Haywood.

Here is the order of waterfalls encountered, starting from the bottom and working upstream:

#1—40-foot-high Stag Brook Falls, a photograph of which can be seen in Den Linnehan's *New York State Splendor* (2008) and *Adirondack Dawn* (2010)

#2—an 8-foot high, nearly vertical cascade that lies just above Stag Brook Falls but is not visible from the base of Stag Brook Falls

#3—a 5-foot-high cascade that drops into a 10-foot-long flume

#4—a 3–4-foot-high cascade that flows into a gentle pool of water

#5—Picnic Table Falls, where a series of cascades drop 15–20 feet into a pool of water overlooked by a picnic table (this spot is too good to pass up if you are planning to stop for lunch or a snack along the way)

#6—Footbridge Falls, where a footbridge crosses a partially formed flume containing several cascades (the main fall, a drop of 6 feet, lies just downstream from the footbridge)

#7—a 4-foot-high cascade

#8—a 15-foot-high cascade

#9—a series of undifferentiated cascades located just below a drainpipe that carries the stream under a ski trail/road

At this point a sign at the junction with the ski trail/road states that you have gone 0.3 mile. Another 0.4 mile remains until you reach the mid-station, farther upstream.

From the other side of the ski trail, continue following the trail upstream, paralleling Stag Brook. There are now blue blazes (old trail markers) as well as new red ones.

#10—a 15-foot-high cascade at the end of a 100-foot-long section of tilted bedrock; in the early spring, this whole section turns into one huge waterfall

#11—Block Falls, where the brook drops 20 feet across an exceptionally broad section of the stream

#12—a 10-foot high cascade

#13—a 15-foot-high cascade

#14—Wrap-around Falls, where the stream drops over a broad, 6-foot-high block that bends slightly into the south bank, almost assuming the "J" shape of a cane

#15—Flume Falls, where the stream narrows, becoming flume-like as it drops over several small cascades

#16—a 25–30-foot-high, inclined cascade topped by a dam

Farther upstream, past where the hiking trail ends, is a 100-foot-long waterslide.

Believe it or not, even by compiling this extensive list I probably have not included every waterfall on Stag Brook. It all boils down to a matter of how one defines a waterfall. For instance, there are many places where it's difficult to say where one waterfall ends and another begins. Even some of the distinctive waterfalls may blur together if there is enough water rushing down in the spring to turn sections of Stag Brook into one big cascade.

To help readers get a better sense of where they are on the trail in relation to the cascades, I have given names to several of the waterfalls based upon their features—Picnic Table Falls, Footbridge Falls, Block Falls, Wrap-around Falls, and Flume Falls.

Footbridge Falls drops into a chasm spanned by a footbridge. Photograph by John Haywood.

To reach the falls: From the ski center follow a dirt road uphill, heading straight up the ski slope. You will have no trouble knowing where to go, for green

218

signs point the way. At 0.1 mile you will come to the clearly marked Stag Brook Falls Trailhead where a narrow sliver of woods with a stream running through it heads uphill between wide ski slopes (44°21.268′N 73°51.790′W).

Picnic Table Falls. Photograph by John Haywood.

To Stag Brook Falls: Sign in at the register and follow the red-blazed trail uphill for 0.05 mile. You will cross over two tiny footbridges at the end and then come immediately out to the base of Stag Brook Falls (44°21.311′N 73°51.853′W). In the spring your view of the waterfall may be compromised by the interfering shape of the gorge and the volume of spray that washes over you as you stand there. In other words, you may not want to get wet in order to see the waterfall head-on. During the drier months of summer and autumn, however, obtaining a good view of the waterfall is not a problem. You can easily cross over the stream at the base of the fall to the opposite rock wall for good views and photos.

To Upper Falls on Stag Brook: From the base of Stag Brook Falls, walk back 50 feet and then turn right onto the upper red-blazed trail. You will immediately come to a lateral view of Stag Brook Falls that provides for excellent photos. From here continue following the upper red-blazed trail as it heads uphill taking you past one cascade after another. By the time that you have gone 0.3 mile, according to the Whiteface Mountain Ski Center's sign, you will reach a ski trail/road that bisects the stream. This is not the end of the Stag Brook Trail, however. Walk across the ski trail/road and then continue along the Stag Brook Trail, heading upstream. The cascades come into view as before, undiminished, one after another. There are blue markers now as well as red ones, the blue ones presumably left over from an earlier time. After another 0.2 mile you will reach a large cascade below a dam. Here the trail leads up to the top of the ravine and out onto the ski slope.

Continue walking uphill along the edge of the woods, going past another ski slope to your right. You will come to a long flat bridge that crosses over the brook. Look upstream from here and you will see a 100-foot waterslide over which one of the ski lifts passes (44°21.488'N 73°52.404'W). I'm not even counting this as one of my fifteen cascades in addition to Stag Brook Falls.

107. CASCADE ON STREAM PARALLELING STAG BROOK

On the opposite side of the main ski slope is an unnamed stream paralleling Stag Brook where a fairly large, torturous gorge contains a 30-foot-high, two-tiered waterfall. A tiny pool has formed at its bottom, from where the stream tumbles over a rock pile to create an additional 8-foot-high cascade.

To reach the falls: From the trailhead for Stag Brook Falls, cross over the ski slope to its north side and follow the first wide ski trail/road east for a couple of hundred feet. There will be a tall wooden fence to your right. Unseen under your feet is a large underground pipe that carries the unnamed stream from one side of the ski trail/road to the other. Once you pass by the fence, look for a path on your right that heads southeast, following along the top of a sloping gorge through a forest of conifers. You will hear the sound of cascading waters wafting up from below. Getting down into the ravine for a closer look is tricky, so it is probably best (and safest) just to enjoy what you can see and hear of the falls without trying to get to the bottom. My recommendation is to scramble down the slope to a large boulder (you will know it when you see it) and do your reconnoitering from there (44°21.307'N 73°51.730'W).

Backtrack to the side ski road and walk across to the other side. You will see an artificial waterfall created by a dam. A path follows past the dam and its millpond and then continues uphill along the stream, but there are no further cascades to be seen.

FALLS NEAR WILMINGTON

Wilmington is a small town along the West Branch of the Ausable River that was settled in 1812 and formed in 1821. At that time it was known as Dansville.

Santa's Workshop, one of the first theme parks for children in the United States, is located in the nearby community of North Pole. It opened in 1949.

Wilmington is best known for its close association with Whiteface Mtn.

108. FLUME FALLS

Three main falls have formed in The Flume, aka the Wilmington Flume, where the West Branch of the Ausable River has cut out a deep, narrow chasm—what one could arguably call a smaller, much shallower version of Ausable Chasm—that extends for nearly 0.2 mile. Donald Morris in the 1989 issue of *Adirondac* sums it up as a "series of narrow drops, totaling 60 vertical feet in 200 yards."

The first waterfall is located just upstream from the Flume Bridge (a sturdy metal bridge built in 1960). The GPS reading here is 44°21.949′N 73°50.501′W. The waterfall stands at a height of 10 feet and contains several worn notches in the bedrock where the stream rushes through during times of normal water flow. Paddlers, I believe, call it the "Depth Charge." This waterfall is framed by Whiteface Mountain in the distance, an uncommonly scenic view even by Adirondack standards and generally one not to be overlooked by photographers. In his 2004 book, *Adirondack Splendor*, Den Linnehan captures the size and majesty of this first fall as viewed from the streambed.

Eighty feet downstream from the bridge is a 15–20-foot-high cascade where the river is momentarily split into two chutes by a large, rocky outcrop (44°21.989′N 73°50.421′W). Potholes can be seen on the west side of the streambed and sidewall by the fall where the greater flow of water is diverted.

The Flume: Upper Fall.

The third cascade, located near the end of The Flume, is called Flume Pool Falls (44°22.031'N 73°50.367'W). I believe that whitewater paddlers call it Demon Falls. Here the river tumbles over an 8-foot-high cascade, momentarily producing a powerful current, as swimmers quickly discover when they try to move against it. The chasm immediately becomes wider here before tapering off and turning again into a regular streambed. A photograph of Flume Pool Falls taken in the 1960s can be seen in the Wilmington Historical Society's 2013 book, *Wilmington and the Whiteface Region: Images of America*.

The Flume was described early on by Seneca Ray Stoddard in his 1874 book, *The Adirondacks Illustrated*: "About two miles south of Wilmington is the natural flume. A long furrow through the rock like the track of a giant plowshare, through which the water shoots like a flash of light." A photo of

The Flume can be seen in Gary A. Randorf's *The Adirondacks: Wild Island of Hope* (2002).

The Flume: Middle Fall.

What makes The Flume so distinctive is the immense amount of exposed bedrock on both sides of the river. There are innumerable places to stand or sit while enjoying views of the river. Both the west-side and east-side paths take you along the full length of the flume. The east-side trail in particular also affords the opportunity to enter into the gorge, allowing access virtually down to the level of the streambed. Best of all, the east and west paths are connected by the Flume Bridge, enabling you to go from one side to the other effortlessly.

On one occasion Barbara and I watched an Outward Bound group using the chasm for cable traverses. The youngsters, suspended high above the raging waters of the gorge, looked as thrilled and excited to make the crossing as we were to watch it.

As always, care should be taken around so much rock and water, particularly if you choose to swim. In 2014 two teenagers died at The Flume while swimming in its turbulent waters. In 2017 a man died at Flume Pool Falls when he was overcome by the powerful current.

To get there: From Lake Placid (junction of Routes 73 & 86) drive northeast on Route 86 for ~9.9 miles.

*To Main (Upper) Parking Ar*ea: As soon as you cross over the Flume Bridge, turn left into a large parking area next to the bridge (44°21.975'N 73°50.466'W).

To west-side trail from main parking area: You have two choices for accessing the west-side trail. The first option is to follow a path downhill from the northeast end of the bridge to the bedrock at the top of The Flume, and then head downstream along the chasm's rocky rim. The second option is to walk northeast down Route 86 for nearly 0.1 mile to the lower parking area, and then follow a trail that heads downhill momentarily and then up to the bedrock overlooking the lower end of The Flume.

To east-side trail from main parking area: Walk across the bridge, turn left, and follow what starts off initially as an old road. From here there are many short spur paths that lead over to views of the chasm. Farther downstream you can even descend to the bottom of the chasm for a more intimate view of the river.

To lower parking area: After driving over the Flume Bridge, proceed north for 0.1 mile and park in a large pull-off on your right (44°22.095'N 73°50.358'W).

To upriver views from main parking area: From the main parking area follow a large, road-like path upriver along the West Branch. In less than 75 feet you will notice an informal path to your left that leads steeply down to the base of the first fall (44°21.949'N 73°50.501'W).

Continuing west on the main, road-like path, you will see to your left the barely visible foundation ruins of an old mill that once operated on the river. Immediately after the foundation ruins you will come to a medium-sized pond (44°21.972'N 73°50.558'W). In the spring the pond is a foot or two higher than the trail, leaking water like a sieve yet contained by a dam created by beavers. It is hard not to be impressed by what these creatures can do. From here you can follow the trail as it parallels the West Branch for another 1.0 mile until it comes out by the Whiteface Mountain Ski Center. There are no cascades to be seen along this stretch of the river, however.

109. WHITEFACE VETERAN'S MEMORIAL HIGHWAY CASCADE

An 8–10-foot-high seasonal roadside cascade composed of a series of ledges can be seen along the left side of the Whiteface Veteran's Memorial Highway as you head up the mountain. This cascade was brought to my attention by photographer John Haywood, who noted it on one of his trips up to the top of Whiteface Mtn.

The chalet-style Whiteface Mountain tollhouse at the start of the summit highway. Postcard c. 1940.

The paved highway to the summit of Whiteface Mtn. (4,867')—the fifth highest peak in the Adirondacks—opened in 1936, seven years after being dedicated by Governor Franklin D. Roosevelt. The road is ~5.0 miles long with an altitude gain of over 2,300 feet. The tollhouse at the beginning of the highway was constructed in 1934 and provides a bit of early-nineteenth-century nostalgia as you drive through and start up the mountain. The tollhouse is located next to Lake Stevens, named after Olympian bobsledder Hubert Stevens.

If you have paid money to access the toll road, after seeing the waterfall you really should drive all the way up to the bare rock summit for panoramic views of the Adirondacks. In fact, the reason for taking the toll road in the first place is to get to the summit. There are spectacular views of the High Peaks as well as Lake Champlain and the Green Mountains of Vermont to the east. This is one of those times when the waterfall is really a secondary attraction. As William Chapman White wrote in *Adirondack Country* (1954), "For those who cannot or do not like to climb, the automobile road up Whiteface, built in 1927 and fought vigorously by most conservation groups, offers a chance to see what the world looks like from an Adirondack peak."

Be aware that Whiteface's exposed summit can get *very* windy, subject to the same stratospheric jet stream that produces the enormously high wind velocities on top of Mount Washington in New Hampshire.

To get there: From Wilmington (junction of Routes 86 & 431) drive west on Route 431 for 2.9 miles. At a fork, continue straight ahead (left) for another 0.2 mile on Route 431 to reach the alpine-style tollhouse and the beginning of the Whiteface Veteran's Memorial Highway (44°24.137'N 73°52.638'W).

From the tollhouse start driving up the road. You will see the fall to your left at ~0.4 mile (44°24.007'N 73°53.146'W). Continue up the mountain for another 4.5 miles to where Route 431 terminates at the "castle" parking area near the summit (44°22.033'N 73°54.329'W).

During the regular season, from mid-June to mid-October, the highway is open daily from 8:45 AM to 5:15 PM. Check their Web site for more specific information.

110. FRENCH'S BROOK FALLS

If you study a High Peaks topographical map, you will clearly see the word "falls" indicated on French's Brook just upstream from where the Schwartz Trail crosses over the brook. I have been to this area three times now and have yet to find the waterfall.

To begin, French's Brook is not a stream to trifle with. It is a fairly substantial watercourse that rises from the shoulder of Whiteface Mtn. above Baldwin Hill and flows northwest into Union Falls Pond. On the first trip that Barbara and I made to French's Brook on April 1, 2000, there was a corduroy bridge that spanned the stream. It allowed us to walk over to the other side of the stream without much effort. You will no longer see it, for the bridge was washed out some time ago either during a spring freshet or during one of the tropical storms.

I call our first hike to French's Brook the April Fool's Day Hike, because we hiked it on April 1 and, indeed, the joke was on us. We expected to see a large waterfall near the trail, but saw only minor cascades.

Before the second hike, I had come to realize that the waterfall must be slightly upstream, just around the corner so to speak and out of sight from where the corduroy bridge earlier spanned the stream. If there was a trail that led to the waterfall or an easy bushwhack, then we could lead our hikers to the fall on one of our Waterfall Weekends. Unfortunately, when Barbara and I got there, there was no faint path on either side leading upstream, nor did it appear that the woods were conducive to an easy bushwhack. Rock-hopping up the stream looked even more problematic. Once again we left empty-handed.

Barbara carefully picks her way across the old corduroy bridge.

On hike number three, which I undertook in 2016, I was determined to find the waterfall even though I knew that I probably would not be able to take others with me to see it. I ended up bushwhacking upstream for perhaps 0.3–0.4 mile, crisscrossing the stream repeatedly and scaling the steep slope up and down through thickets and dense pine brush. It was probably the most demanding bushwhack I have undertaken. Could I have missed the waterfall? Yes, it's possible, since I sometimes was hiking on the upper slope trying to listen for the sound of a cascade below. But maybe the waterfall listed on the topographical map was incorrectly placed and I simply just didn't go far enough upstream. Your guess is as good as mine. At this point all I can tell you is that the map shows a waterfall on French's Brook close to the Schwartz Trail.

To be sure, a number of small cascades are visible near the stream crossing, but none are higher than 3–4 feet. Whether this makes the trek worthwhile is something that you will have to decide for yourself. One thing is for sure—it is a pleasant hike and an easy one, and you are likely to have the trail all to yourself.

Small cascades on French's Brook.

In *Discover the Northern Adirondacks* (1988), Barbara McMartin mentions that the Schwartz Trail was once part of a former road that went from Little Montreal (a makeshift village settled by French Canadian loggers) toward Lake Placid. Sandra Weber, in *The Lure of Esther Mountain* (1995), recounts a story by Jim Goodwin that took place on this road in 1931. According to Goodwin: "I had managed to drive my Model A Ford to the top of the notch on the old Wilmington–Franklin Falls road ... [then] I followed that wood road [Schwartz Trail] to the French's trail and climbed it to the top of Whiteface, descending the same way."

To get there: From Wilmington (junction of Routes 86 & 431) drive west on Route 431 for 2.9 miles. At a fork where Route 431 heads straight ahead to the summit of Whiteface Mtn., bear right, continuing now on Route 18 for another 0.6 mile (or exactly 0.1 mile past the trailhead for the "Copper Kiln Pond Lean-to/Bonnieview Road," which you will see after 0.5 mile on your right). The unmarked trailhead for the Schwartz Trail will be to your left, quite obvious once you realize that it is there, but easy to miss initially. Turn around and park next to it (approximately 44°24.347'N 73°53.142'W).

Follow the Schwartz Trail south for a total of 2.3 miles. The trail, an abandoned logging road, is wide and in great condition with very little change in elevation. It is the perfect trail for a leisurely trek as it wends its way around the coattails of Esther Mtn. (4,239'), supposedly the only peak in the Adirondacks named after a woman—15-year old Esther McComb, who climbed it in 1839. At about 0.05 mile before the end of the trail, the river suddenly comes into view as it rumbles along in a deep gorge below. Continue until you come to a point where the road/trail suddenly contracts into a narrow path and then disappears. Turn around, backtrack fifty feet, and you will end up at the point where the trail previously crossed over the stream via a crude bridge. If you look closely you will see the old road continuing on the opposite side of French's Brook. Keep in mind, though, that this is not an easy stream to cross. Even after months of relatively dry weather, there was still a significant amount of water flowing when I arrived on my third visit, making it somewhat challenging to cross to the opposite side.

There are small cascades in the stream, but they are difficult to get to.

Based upon what is shown on the topographical map, the main falls are no more than 0.3 mile upstream at a GPS reading of around 44°23.311'N 73°54.851'W—or maybe they're not!

If you wish to bushwhack upstream to look for these falls, be my guest. Just remember that from this point on the hike changes from the Schwartz Trail to the "Schwarzenegger Trail!"

Postscript

Since 2003 I have authored four regional waterfall guidebooks to Eastern New York—*Adirondack Waterfall Guide, Catskill Region Waterfall Guide, Hudson Valley Waterfall Guide,* and *Mohawk Region Waterfall Guide.* I have also witnessed the genesis of multiple New York State waterfall Web sites, three New York State waterfall guidebooks featuring some of the state's most prominent waterfalls, two regional guidebooks to waterfalls of Central and Western New York State, and a number of waterfall calendars, with one series featuring the Adirondacks. Waterfall meet-up groups have formed, and hiking groups now routinely lead treks to waterfalls or include them along trips to other destinations.

Additionally, attempts have been made over the years on at least three separate occasions to systematically catalogue all of the waterfalls in New York State (a nearly impossible task, but yet one worthy of being undertaken).

One of the newest waterfall-related projects to arise is the Hamilton County Waterfall Challenge, where hikers can earn a patch by visiting a combination of the 19 waterfall sites listed in the county.

In 2016 new heights were aspired to when John Haywood of John Haywood Photography, Edward Smathers, creator of the Dig the Falls Web site, and Bobbie Sweeting, creator of the Bobbies Waterfalls Web site, formed the New York Waterfall Conservancy, their mission being to promote the beautification of New York waterfalls and to ensure their preservation for future generations to enjoy.

Waterfalls, it seems, have come into their own. May they always call to us.

Acknowledgments

Many thanks go to: John Haywood, eminent photographer, waterfall hunter, publisher, and all-around good friend, for his many contributions to this book; to Bradley Knapp, Plant Operator for Eagle Creek Renewable Energy, for his contributions to the chapters on Alice Falls and Indian Falls; to Mike Sheridan, Manager, Elk Lake Lodge, who provided helpful information on Elk Lake's history and reviewed the write-up for accuracy; to Kathryn Reiss, for reviewing the write-up on High Falls Gorge and for comments made; to Amy & Chris Bergman, creators of the blog ItsMoreFunOutdoors.com, for photographs of the Trap Dike, Fall on Tributary to Lake Colden, and MacIntyre Falls; to John Rowen, who started me thinking about the Beaver Deceiver; to Andrea Anesi, Archivist/Librarian, Essex County Historical Society Research; to Loren G. Dobert, for valuable information on the hike to the Boquet Canyon and for his photo contribution; to John Holmes, for information provided on Fairy Ladder Falls and for his photo contributions; to the New York State Library, where I have sat for hours doing research; to the Keene Valley Library, a wonderful resource; to Neal Burdick for adding immeasurably to the book with fresh and different things to say about waterfalls in his foreword; to Barbara Delaney, my wife and fellow author, who accompanied me to many of the waterfalls listed, provided valuable input, and who also proofread the manuscript; and last, but not least, Steve Hoare, publisher of Black Dome Press, which has indisputably published more waterfall guidebooks than anyone else in the world.

Rocky Falls

About the Author

Licensed New York State hiking guide Russell Dunn is the author of twelve previous popular and critically acclaimed guidebooks to eastern New York State and Massachusetts:

Adirondack Waterfall Guide: New York's Cool Cascades (Black Dome Press, 2003)

Catskill Region Waterfall Guide: Cool Cascades of the Catskills & Shawangunks (Black Dome Press, 2004)

Hudson Valley Waterfall Guide: From Saratoga and the Capital Region to the Highlands and Palisades (Black Dome Press, 2005)

Mohawk Region Waterfall Guide: From the Capital District to Cooperstown and Syracuse (Black Dome Press, 2007)

Berkshire Region Waterfall Guide: Cool Cascades of the Berkshires & Taconics (Black Dome Press, 2008)

A Kayaker's Guide to Lake George, the Saratoga Region & Great Sacandaga Lake (Black Dome Press, 2012)

A Kayaker's Guide to New York's Capital Region: The Hudson & Mohawk Rivers from Catskill & Hudson to Mechanicville, Cohoes to Amsterdam (Black Dome Press, 2010)

Paddling the Quiet Waters of Mid-Eastern New York (Troy Book Makers, 2014)

Penultimate Paddles: Southeast Adirondacks (John Haywood, 2016)

Rockachusetts: An Explorer's Guide to Amazing Boulders of Massachusetts (Dunn & Butler, 2016), coauthored with Christy Butler

Trails with Tales: History Hikes through the Capital Region, Saratoga, Berkshires, Catskills and Hudson Valley (Black Dome Press, 2006), coauthored with Barbara Delaney

Adirondack Trails with Tales: History Hikes through the Adirondack Park and the Lake George, Lake Champlain & Mohawk Valley Regions (Black Dome Press, 2009), coauthored with Barbara Delaney

Russell Dunn is also the author of:

Adventures around the Great Sacandaga Lake: Revised & Illustrated (Troy Book Makers, 2011)

Connecticut Waterfalls: A Guide (Countryman Press, 2013), coauthored with Christy Butler

Vermont Waterfalls: A Guide (Countryman Press, 2015)

Ausable Chasm: In Pictures and Story (John Haywood Photography, 2015), coauthored with John Haywood and Sean Reines

3-D Guide to the Empire State Plaza and Its Large Works of Art (Troy Book Makers, 2012), coauthored with Barbara Delaney

Ausable Chasm in 3-D (John Haywood Photography, 2015), coauthored with John Haywood

Natural Stone Bridge & Caves in 3-D (John Haywood Photography, 2016), coauthored with John Haywood

Ogunquit in 3-D (John Haywood Photography, 2016)

Adirondacks in 3-D (John Haywood Photography, 2016)

Helderbergs in 3-D (John Haywood Photography, 2016)

Great Sacandaga Lake in 3-D (John Haywood Photography, 2016)

Albany in 3-D (John Haywood Photography, 2016)

Catskills & Shawangunks in 3-D (John Haywood Photography, 2017)

Berkshires in 3-D (John Haywood Photography, 2017)

Rome in 3-D (John Haywood Photography, 2017)

Florence & Scenes of Italy in 3-D (John Haywood Photography, 2017)

Adirondack Waterfalls in 3-D. Volume One (John Haywood Photography, 2017)

Adirondack Waterfalls in 3-D. Volume Two (John Haywood Photography, 2017)

Together with Barbara Delaney (also a NYS-licensed guide), Dunn leads hikes to waterfalls in the Adirondacks, Catskills, and Hudson Valley, as well as to other sites of exceptional beauty and historical uniqueness. Dunn and Delaney give frequent lecture & slideshow presentations to regional historical societies, libraries, museums, civic groups, organizations, and hiking clubs on a wide variety of subjects, ranging from waterfalls to natural areas of historic interest. Dunn can be reached at rdunnwaterfalls@yahoo.com.

Russell Dunn

Index

A

Adgate Falls, 159
Adgate, Matthew, 159
Adirondack Canoe Waters: North Flow, 32, 34, 149, 151, 153, 184, 204
Adirondack Country, 226
Adirondack Dawn, 217
Adirondack Days, 169
Adirondack Guide, The, xxviii, 83
Adirondack High: Images of America, 72
Adirondack High: Images of America's First Wilderness, 211
Adirondack Landscape, The, 87, 185
Adirondack Land Trust, 32
Adirondack Life Magazine, xxvii, 3, 6, 11, 74, 150, 182
Adirondack Life 1999 Annual Guide to the Adirondacks, xxviii, 43
Adirondack Life's 2007 Annual Guide to the Great Outdoors, 130
Adirondack Loj, 171, 172
Adirondack Moments, 96, 168
Adirondack Mountain Club, 21, 133, 145, 172, 190
Adirondack Mountain Reserve, 74–77, 122, 136
Adirondack Northway, 37, 43, 45, 46, 48, 49, 170
Adirondack Park, xxviii
Adirondack Pass, 180
Adirondack Peak Experiences, 185
Adirondacks: A Special World, The, 17, 139
Adirondacks Forever Wild, The, 125, 167, 180
Adirondacks Illustrated, The, 5, 72, 85, 90, 117, 159
Adirondacks: In Celebration of the Seasons, The, 6
Adirondack Ski Touring Council, 165
Adirondack Splendor, 6, 82, 87, 150, 168, 176, 197, 208, 211, 221
Adirondacks: Postcard History Series, The, 26, 122
Adirondacks, The, 78, 82, 86, 87, 125, 135, 167, 183
Adirondacks: Views of an American Wilderness, 83
Adirondacks: Wild Island of Hope, The, 87, 139, 149, 223
Adirondack Trail Improvement Society (ATIS), 75, 112, 115
Adirondack Trails: High Peaks Region, 21, 179, 184
Adirondack Venture: Images of America, 192
Adirondack Waterfall Guide, 77, 106, 123, 130
"Adirondack Waterfalls", xxviii
Adirondack Waters: Spirit of the Mountains, 17, 73, 114
Adirondack Waterways: 2001 Collectors Issue, 74
Adirondack Wilderness, 40, 87
Adirondac Magazine, 44, 77, 204, 212, 221
Adventure Trail, 158
Agassiz, Louis, 152
Albany County, xxv
Algonquin Brook, 187
Algonquin Peak, xxvii, 179, 180, 187
Alice Falls, 147, 153, 154–156
Alice Falls Hydro, 147, 154
Alice Falls Pulp Mill, 156
Altschuler, David, 4
Ancient Adirondacks, The, 183
Anderson Falls, 147, 151–153, 154
Anderson, John, 151
Apperson, John, 170, 181
Armstrong Mountain, 75, 84

Army Corps of Engineers, 166
Around Keeseville: Images of America, 163
Artist Falls, 94–96
Ash Craft Brook, 48
Ash Craft Brook Falls, 48, 49
Ash Craft Pond, 48
Atlantic Monthly, 117
Atlantic Ocean, 184
Atlantic salmon, 30
At the Mercy of the Mountains, 59, 72, 186
Ausable Chasm, xxiv, 26, 85, 86, 157, 158, 164, 210, 221
Ausable Chasm Horsenail Works, 159
Ausable Chasm in Pictures and Story, 158
Ausable Club, 75, 78, 86, 89, 90
Ausable Forks, 76, 147, 149, 151, 200
Au Sable Horsenail Factory, 152, 153
Ausable Lake, Lower, 74–76, 78, 79, 87–89, 128
Ausable Lake, Upper, 75, 76, 88
Ausable Point, 200
Ausable River, 74, 76, 153, 156, 158, 184, 191, 200
Ausable River, East Branch, 70, 74–79, 84, 89, 90, 105–107, 112, 114, 115, 117, 119, 121, 122, 124, 125, 127, 147–149, 151, 164, 167, 200
Ausable River, East Branch Gorge, 80
Ausable River, Little, 163
Ausable River, West Branch, 74, 76, 147, 149, 160, 176, 191, 199, 200, 204, 205, 207–209, 211, 212, 216, 221, 224
Austin Glen, xxxiv
Avalanche Lake, 183–186
Avalanche Pass, 182–184, 186
Avalanche Pass Falls, 182–184
Avery Bridge, 191

Avery, Simeon Shipman, 191

B

Baldwin Hill, 226
Bark Eater Inn, 166
Barnett, Lincoln, 183
Barry, John D., 191
Bartlett Falls, 24
Barton Brook, 27
Basin Mountain, 75, 88
Battaglia, Nancy, 107
Beaver Deceiver, 117
Beaver Meadow Bridge, 82, 84, 89
Beaver Meadow Brook, 82
Beaver Meadow Falls, xxviii, 43, 81–84, 87, 90
Beaver Meadow Farm, 6
Beaver, Tony, 107
Beede Brook, 60–62
Beede Brook Falls, 60–65
Beede Hill, 122
Beede House (hotel), 75
Beede, Orlando, 73
Beede, Smith, 72, 75,
Beer Bridge Way, 69, 107
Beer Walls, 67
Beer Walls Falls, 67–69
Beginning…Wadhams, 1820–1993, 31
Beisel Jr., Richard H., 160
Bennies Brook, 138, 139, 142
Benson, 191
Berkshire Region Waterfall Guide, 144
Berkshires, xxv
Berlin Iron Company, 152, 153
Bernays, David, 86
"big blowdown of 1950", 202
Big Slide Mountain, 132, 133, 146
Birmingham Falls, 155, 159, 162
Bishop, Basil, 6
Black Dome Press, 77, 107, 123
Black River, 38
Blake Mills, 101

Blake Peak, 39, 75
Block Falls, 218
Blueberry Falls, 121
Blueberry Mountain, 121, 124
Blue Ledge, 39
Blue Ledge Falls, 39
Blue Lines: An Adirondack Ice Climber's Guide, 65
Blue Mountain, 119
Blue Ridge Falls, 38, 39
Blue Ridge Falls Camp Site, 38
Board of Geographic Names, 2
Boas, Keith, 8, 17, 31, 83, 149, 207, 211, 212
Bobbies Waterfalls, 230
Boquet Canyon, 14–16
Boquet Canyon Falls, 14–16
Boquet, Charles, 2
Boquet, General, 2
Boquet, Little, 23–25, 28
Boquet River, xxviii, 2, 3, 5, 7, 8, 14, 17, 20, 22–24, 32–34
Boquet River Association (BRASS), 2
Boquet River Falls, 11
Boquet River, North Fork, 2, 8, 9–11, 16, 19, 21, 52, 56
Boquet River, South Fork, 2, 9, 20, 21
Boundary Brook, 187
Boundary Peak, 187
Bouquet, 34
Bowie, Mark, 6, 17, 71, 73, 107, 114
Boxcar Swimming Hole, 12
Boyoma Falls, 161
Braman, Jessie, 30
Braman's Mills, 30
Branch, The, 23, 38, 39
Bright, George, 84
British, 2, 41
Brockett, L. P., 104
Brodhead, Charles, 60
Bronski, Peter, 59, 72, 186
Brown, Deacon Levi, 25

Brown, John, 165
Brown, Mary, 165
Brown, Scott, 89, 96, 111
Bullock Dam, 89
Burdick, Neal, xv, xvi, 130
Burlington, 125
Burnside, James. R., 66, 80, 85, 178, 180, 183
Burr Map, 148
Bushnell Falls, 128, 135, 136, 176
Bushnell, Rev. Horace, 135
Bushwhacker's View of the Adirondacks, A, 20, 43, 101, 102
Butler, Christy, xxxiv, xxxv, 202
Buttermilk Falls, 86, 163
By Foot in the Adirondacks, 70, 168, 171

C

Camp Baco, 3, 4
Camp Grace, 134
Camp Thistle-Do, 134
Canadian Pacific Railway Police Service, 33
Canadians, xxvii, 228
Carthage Road, 38
Cascade Brook, 84, 87, 168
Cascade Lake Falls, xxiv, 168–171
Cascade Lake House, 169, 170
Cascade Lakes, xxiv, 165, 168
Cascade Mountain, 168
Cascade on Twin Pond Outlet Stream, 10, 11
Cascade Pass, 165, 168
Cascades along East Trail to Giant Mountain, 22
Cascadeville, 169
Casilear, John, xxxi, 76
Cathedral Rocks, 78
Catskill Mountain Region Guide, Xxxii

Catskill Region Waterfall Guide, 230
Catskills, xxv, xxxi, xxxiv,
Caution: Safety Tips, xvii–xxii
caves, xxxiii, 111
Cedar Point Road, 41
Champagne Falls, 125, 126
Champlain Fibre Company, 35
Champlain Valley, 2, 76
Chapel Pond, 50, 58, 60, 65–67, 105
Chapel Pond Canyon, 67
Chapel Pond Falls, 65–67
Chapel Pond Pass, 59
Chapel Pond Slab Cascade, 59, 60
Chase, Greenleaf, xxviii
Chasm Cascade, xix, xxviii, 17–19
Chicken Coop Brook, 136
Chrissie, 77
Chubb, Joseph, 191
Chubb River, 191, 192, 197, 199
Chubb River Chasm, 192, 193
Chubb River Falls, 193–195, 197
Civilian Conservation Corps (CCC), 181
Clarke, Robert, 185
Clark, Herbert, 77
Clarksville, xxxiii
Clarksville Cave, xxxiii
Clear Lake, 171
Clear Pond, 41
Clifford Brook, 167
Clifford Brook Falls, 167, 168
Climax Falls, 211
Clinton County, 23, 155
Clough Brook, 117
Clough, Esther, 117
Clough, Parley, 117
Coates, John, 30
Coate's Mill, 30
Cohen, Alan, 4
Colden Dam, 189
Colden, David C., 185

Colden, Mount, 181, 183–185
Coleman, Samuel, xxxi, 76
Colvin, Mount, 75, 99
Colvin Range, 75
Colvin, Verplanck, 40, 95, 101, 180
Coney Island, 184
Connery Pond, 202
Conservation for Public Recreation, 44
Conservationist, New York State, xxxii
Cooper's Falls, 163
Copper Kiln Pond, 228
"Creek Freaks", 3
Crossing the River: Historic Bridges of the Ausable River, 156
Crowfoot Brook, 46, 47
Crowfoot Brook Cascades, 46–48
Crowfoot Pond, 46, 47
Crown Point, 38
Crystal Falls, 180

D

Dansville, 221
Day Trips with a Splash, 12, 31, 142
Deadwater Pond, 46
Deer Brook, 109, 111, 112
Deer Brook Cascades, 129, 130
Deer Brook Falls (Johns Brook), 129, 130
Deer Brook Falls (Route 73), 109–112
Deer Brook Gorge, 109–111
Deerfield River, 34
Deer Mountain, 5
Demon Falls, 222
de Muth, Otto, 6
Denton Pond, 26, 27
Department of Environmental Conservation (DEC), 131, 181, 191
Department of Transportation (DOT), 127

depth charge, 221
Desormo, Maitland C., 148
Devil's Oven, 157
Dial Mountain, 75
Dibble, Seth, 128
Dig the Falls, 230
Dipper Brook, 60
Dipper Pond, 60, 65
Discover the High Peaks, 101, 132, 187
Discover the Northeastern Adirondacks, 28, 62, 228
Dix Mountain, 2, 13
Dix Mountain Wilderness, 42
DNA, xxix
Dobert, Loren G., 14
Doeffinger, Derek, 8, 17, 31, 83, 149, 207, 211, 212
Doll, Poncho, 12, 131, 142
Duggan, Mike, 122
Durand, Asher B., xxxi, 76

E

East Berlin, 153
East Dix, 2
East Hill, 104
East River Trail, 89–103
Eddy Forge, 25
Edmund, 169
Edmund Pond, 169
Elba Iron & Steel Manufacturing Company, 179
Elba Iron Works, 191
Elephant Head, 157
Elizabethtown, 2, 21, 24, 26–28
Elizabethtown, NY: Bicentennial Celebration, 1798–1998, 25
Elk Lake, 38, 39, 41
Elk Lake–Clear Pond Forest Preserve, 40
Elk Lake Lodge, 40
Elk Pass, 91
Elm Tree Inn, Monty's, 164
Emerson, Ralph Waldo, 136, 152

Emmons, Ebenezer, xxxi, 185
Emperor Slab, 59
Engelhart, Steven, 156
Ensminger, Scott A., 87, 150
Erick's Camp, 43
Erie Canal, 6
Essex County, 23, 104, 155
Essex County Fish Hatchery, 50
Estes Cemetery, 119
Esther Mountain, 229
E-Town, 26
Euba Mills, 9
Exploring the Adirondack Mountains 100 Years Ago, 155
Exploring the 46 Adirondack High Peaks, 66, 80, 85, 178, 180, 183

F

Fairy Ladder Falls, 101–103, 179
Falls Brook, 24
Falls, The (Wadhams), 30
Farb, Nathan, 82, 87, 106, 107, 124, 125, 135, 149, 167
Felt, Aaron, 30
Fenn, Harry, 156
Fern Gully, 9
"finest square mile", 171
Fingerbowl Pond, 60
Fish & Wildlife, U.S., 128
Fitch, John, xxxi, 76
Five-Star Trails in the Adirondacks, 87
Flammer, Edward, 26
Fletcher Allen Health Center, 125
Flowed Lands, 184, 189
Flume Bridge, 221, 223
Flume Brook, 114, 115
Flume Cottage, 114
Flume Falls (Chasm Falls), 17
Flume Falls (Flume Brook), 114, 115
Flume Falls (Stag Brook), 218

Flume Falls (Wilmington), 220–224
Flume Pool Falls, 222, 229
Flume, The (AMR), 93, 94
Flume, The (Wilmington), xxiv
Foley, Matt, 29
Follensby Pond, 152
Footbridge Falls, 217
Forever Wild, 181
Forever Wild: The Adirondacks, 79, 83, 139
Forty Sixers, 77
French, 2, 200
French & Indian War, 2
French's Brook, 226, 227
French's Brook Falls, 226–229
"freshet of 1856", 76
From Niagara to Montauk, 210
Frontier Town, 37, 42
Funnel Falls, 141

G

Gallos, Phil, xxxi, 70, 168, 171
Garden, The, xxiv, 128, 129, 137, 146
"Gateway to the Adirondacks", 37
"Gateway to the Olympics and High Peaks", 151
General Electric, 181
Geological History of the State of New York, A, 104
Geology of the Adirondack High Peaks, 67, 82, 86, 93, 107, 122, 179, 185
Georgia-Pacific Corporation, 35
Giant of the Valley, 60
Giant Mountain, 50, 56, 58–60, 70, 73, 110
Giant Mountain Wilderness, 22, 70
Giants Dipper, 60
Giants Nubble, 65
Giant's Washbowl, 50, 60
giardiasis, xviii
Gibbs, John, 127

Gifford, Stanford, xxxi, 76
Gilbaldi, Paul L., 207
Gill Brook, 91–103
Gill Brook Falls, Upper, 99–101
Gill Brook Steps, 94
Gilliland, Elizabeth Phagan, 26
Gilliland Park, 36
Gilliland, William, 26, 35
Glendale Falls, 144
Glen, The, 148
Goff, Dr. Alphonso, 119
Goodwin, Tony, 184, 228
Google Earth, xxvi, 173, 175
Gothics, 75, 79, 83, 84, 88
"Grand Canyon of the East", 157
Grand Canyon Skywalk, 211
Grasse River, South Branch, 87
Graves, David, 125
Great Falls, 155
Great Lakes, xxxi
Great Range, 73, 75, 78, 79, 110
Greek, xxix
Green Mountains (Vermont), 226
Griffiths, David, xxiv
Guide to Adirondack Trails: High Peaks Region, 184

H

Halcomb, Jean, 44
Hale, David, 76
Hamilton County Waterfall Challenge, 230
Hammond Pond Wild Forest, 45, 46
Hanging Spear Falls, xxviii, 189
Harper's New Monthly Magazine, 169
Harris, Barbara, 176
Hawaii, 196
Haystack Mountain, Little, 75, 127
Haystack, Mount, 75
Haywood, John, xxxv, xxxvi,

239

56, 107, 158, 225
Headley, J. T., xxxi
Heald, David, 76, 127
Healy, Bill, 17, 118, 139
Heart Lake, 171, 176
Heaven up-h'sted-ness!, 70, 176
Hedgehog Mountain, 75, 79, 109, 114
Heilman II, Carl, 83, 107
Heraclitus, xxix
Heydays of the Adirondacks, The, 148
Hidden Cascade, 51, 52
Highby, Levi, 35
High Falls, 86
High Falls Gorge, xxiv, 210, 211, 216
High Falls Gorge Falls, 200, 210, 211
High Peaks, xxiv, xxvii, xxxi, 2, 37, 44, 60, 67, 75, 104, 128, 131, 133, 134, 137, 171, 226
High Peaks: A History of Hiking the Adirondacks from Noah to Neoprene, 119
High Peaks Information Center, 172, 174
High Peaks of Essex: The Adirondack Mountains of Orson Schofield Phelps, The, 118
"Hiking the Waterfall Trail", 77
"Hiking Waterfalls in New York", 6, 213
Hill, Governor David B., 206
History of Clinton and Franklin Counties, New York, 159
History of Essex County, 39
History of Wadhams, A, 31
"hitch-up Matilda", 186
Hoffman, Charles Fenno, xxxi
Holcomb Mountain, 5
Holcomb Pond, 204
Holcomb Pond Cascade, 204
"Holy Waters", 11, 150
Homer, Winslow, xxxi, 76, 119

Hoosac Tunnel, 33, 34
Hopkins Brook, 115, 116
Hopkins Brook Falls, 115–117
Hopkins Mountain, 105, 115, 117
Hopkins, Rev. Erastus, 115
Horrell, Jeffrey L., 159
Horseshoe Falls, 147, 161, 162
Hotel Douglas, 163
Houck, Frank, xxviii
Howard Mountain, 176
Hoyle, Robert, 152
Hudson River, xxxi, 37, 184
Hudson Valley Waterfall Guide, 230
Hull, Alden, 122
Hull Basin Brook, 124
Hull Basin Brook Falls, 124, 125
Hull, Eli, 122
Hull, Joseph, 122
Hulls Falls, 75, 90, 122–124, 147
Hulls Falls House, 122
"Hunting for Waterfalls", xxix
Huntington Gorge, 4
Hurd, Duane Hamilton, 159
Hurricane Mountain, 28
Hydes Cave, 157
hydropower, xxxii

I

ice climbing, 65, 73, 86
Ildan, Mehmet Murat, 3
Indian Falls (Ausable River), 153, 154
Indian Falls (Marcy Brook), xxii, xxxi, 176, 179–182
Indian Head, 98
Indian Lake, 180
Indian Pass, 176, 180
Indian Pass Brook, 176, 180
Indian Pass Creek, 180
Indian Point, 180
Indian River, 180
Industrial Revolution, xxxii
Inner Sanctum Trail, 157, 162

In Stoddard's Footsteps, 71
Interior Outpost Ranger
  Station, 128, 131–133, 137, 138, 143, 146, 188
*International Waterfall
  Classification System*, 160
In the Beginning....Wadhams, 31
In the Heart of the Mountains, 105
Irene, Tropical Storm, 74, 126, 127, 181
Iroquois Peak, 180

J
Jack Rabbit (Ski) Trail, 165
Jack's Dam, 191
Jackson Brook, 24
Jacob's Well, 157
Jamieson, Paul, 32, 34, 149, 151, 153, 164, 184, 185, 204
Jay, 76, 147, 151, 200
Jay Covered Bridge, 150
Jay Falls, 149–151
Jay, John, 151
Jay Mountain, 148
Jay Mountain Wilderness Area, 24
Jay, Upper, 76, 151
Johannsen, Herman "Jack Rabbit", 165, 213
John Haywood Photography, 230
Johns Brook, 127–146
Johns Brook Bridge, 127
Johns Brook Lodge, 128, 133, 134
Johns Brook, Lower Flume, 143, 144
Johns Brook Trail, 128
Johns Brook, Upper Flume, 131, 132

K
Kaatskill Life, xxix, 5

Keene, 76. 90, 122, 147, 164, 165, 168
Keene Flats, 104
Keene Heights Hotel Company, 75
Keene Valley, xxiv, xxvi, 69, 76, 104, 105, 115, 117, 147
Keene Valley, Greater, xxiv
Keene Volunteer Rescue Squad, 125
Keese, Richard, 152
Keeseville, xxiv, 74, 76, 147, 152, 164, 210
Kenset, John F., xxxi, 76
Kent, Israel, 125
kettle lake, 71
keystone arch bridge, 152
King Slab, 59
Kirschenbaum, Howard, xxviii, 83
Klondike Brook, 173, 175
Klondike Brook Falls, 173–176
Klondike Brook Notch, 173
Klondike Dam Camp Lean-to, 175, 176
Klondike Trail Waterslide, 175
Knob Lock Mountain, 24
Kozma, Ethel L., 31
Kraus, James, 72, 96, 168

L
Lake Champlain, xxxi, 30, 34, 74, 163, 184, 200, 226
Lake Colden, 131, 184, 187
Lake Keene, 50, 76, 104
Lake Placid, xxiv, 164, 199, 200, 206, 214
Lake Placid Airport, 191
Lake Placid Club, 170, 171
Lake Placid Company, 171
Lake Placid Ski Council, 213
Lake Road, 74, 77, 90
Lake Stevens, 225
Lampson Falls, xxviii
Land & Forest Division, DEC's, 131

241

Langmuir, Irving, 181
Laphams Mills, 163
Lawrence Jr., Richard W., 6
Lawrie, Alexander, 91
Lindsay Brook, 44, 46
Lindsay Brook Falls, 44–46
Lindteigen, Sue, xxiv
Linnehan, Den, 6, 82, 87, 107, 150, 168, 176, 197, 208, 211, 217, 221
"Liquid Assets", xxviii
Little Boquet, 23–25, 28
Little Falls (Merriam Forge), 32
Little Montreal, 228
Little Trout Brook, 163
Lobdell, John, 25
Lobdell, Lewis, 25
log slip, 148
"Long Chute, The", 152
Long Pond, 169
Longstreth, T. Morris, 70
Long Trail, 190
Lost Lookout, 84, 87
Lowell, Russell, 152
Luckhurst, Neil, 185
Ludlum, Stuart D., 155
lumbering, xxxii, 37, 134, 176, 180
Lure of Esther Mountain, The, 228
lyme disease, xxii

M

MacIntyre, Archibald, 179
MacIntyre Brook, 178, 184
MacIntyre Brook Falls, 178
MacIntyre Falls, 178, 179
MacIntyre Iron Works, 185
MacIntyre Range, 180, 184, 199
Mackenzie, Mary, 165
Macomb, General Alexander, 41
Macomb Mountain, xxviii, 41, 43
Main Fall, 211

Malfunction Junction, 2
Mallory, Nathaniel, 151
Mallory's Bush, 151
Marble Mountain, 213
Marcy Air Field, 119, 121
Marcy Brook, 179–181, 199
Marcy Dam, 131, 181, 184–187
Marcy Mountain, Little, 127, 179
Marcy Pond, 181
Marshall, George, 77
Marshall, Robert, 77
Massachusetts, 144
Mather, J. H., 104
McClellan, Katherine Elizabeth, 105
McComb, Esther, 229
McEntee, Jervis, xxxi
McIntyre, Archibald, 191
McMartin, Barbara, xxviii, 28, 43, 44, 62, 101, 132, 187, 228
Mellor, Don, 65, 67
Meriam Forge Falls, 32–34
Meriam, P. D., 32
Meriam, William P., 32
Military and Civil History of the County of Essex, New York, 76, 84, 159, 162, 168, 200
Miller, Edna, 169
Miller, Nicanor, 169
Milltown, 35
Milltown Falls, 34
Minerva, 3
Minetor, Randi, 6, 213
Mineville, 29
mining, 23
Mini-Trap Dike, 59, 60
Mitchell's Cave, 63
Mohawk Region Waterfall Guide, 230
Monument Falls, 200, 205–207, 216
Moriah, 37
Morris, Donald, 3, 32, 34, 149, 151, 153, 184, 204, 212, 221

Mossy Cascade, 69, 105–108
Mossy Cascade Brook, 105, 109
Mountaineer, The, 127
Mount Marcy, 28, 179, 181, 186, 199, 200
Mount Marcy: The High Peak of New York, 180
Mount Washington, 226
Mr. Van Ski Trail, 175
Mud Pond, 40
Murray, William Henry Harrison, xxxi
Myrick, Barnabus, 31

N

Nash, Roderick, xxxi
Native Americans, xxxi, 6, 180
Natural Stone Bridge & Caves, 210
Nedele, Jon, xxviii
Neilson, William G., 75
Newport Brook, 47
New Russia, 6, 23, 115, 123
Newton, Sir Isaac, xxvii
New York Exposed: The Whitewater State, 122, 128
New York State Library, xxvi
New York State Splendor, 87, 217
New York Waterfall Conservancy, 230
New York Waterfalls, 81, 96, 111
Niagara Brook, 44
Niagara Falls, 44, 87, 160
Niagara Mohawk, 29
Nichols Brook, xxviii, 164
Nichols Brook Falls, 163, 164
Nippletop Mountain, 39, 75, 99
Noble Mountain, 18
Noonmark Mountain, 73, 75, 77, 144
North Elba, 165, 168, 179, 191
North Fork Gorge Cascades, 8–10
North Hudson, xxiv, 37–39
North Pole, 221
Northside Trail, 128

Northville, 191
Northville-Placid Trail, 190–198
Northville-Placid Trail, 196
Northwest Bay Hopkinton Road, 165
Northwest Bay Trail, 164
Nowicki, Richard, 196
Nye, Bill, 180

O

Of the Summits of the Forests: ADK 46-R, 170
O.K. Slip Falls, xxviii
Old Military Road, 165
Old Military Tract, 60,
Old Tavern House, 6
One Hundred Views of the Adirondacks, 125, 135, 149
Opalescent Falls, 189
Opalescent River, 184, 189
"Other Niagara Falls, The", xxviii, 43
Otis, J. Henry, 169
Outside Magazine, 40
Outward Bound, 223
Owen Pond, 208, 209
Owen Pond's Outlet Cascade, 208, 209

P

Page, Kyle M., 163
Parker, Cecil, 121
Parker, John Adams, xxxi, 76
Parmerter, Jacob, 39
Parton, Arthur, xxxi, 76
Payne, Daniel F., 31
Peck, Charles, 101
Perkins, Frederick, xxxi, 76
Perkins, T. S., xxxi, 76
Phelps Brook, 117, 119
Phelps, Ed, 128
Phelps Falls, 117–119
Phelps Mountain, 119
Phelps, Orson Schofield (Old

Mountain) Phelps, xxvii, 90, 101, 117, 119, 128
Phelps Trail, 128, 129–136, 137
Picnic Table Falls, 217
Pilcher, Edith, 83, 88, 94
Pillar, The, xxxv
Pinnacle Mountain, 75
Pitchoff Mountain, 165, 168
Plattsburg, 41
Porter, Douglas, 163
Porter, Eliot, 79, 83, 139
Porter Mountain, 119
Positive Reinforcement, 67
Pottersville, 210
"Power of Waterfalls, The", xxxii
Prescott & Sons, R., 153
"primitive man", 117
Putnam Brook, 70, 72
Putnam, H. A., 23
Pyramid Brook, 79
Pyramid Falls, 79, 80

Q
Quarry Pool, 205
Quarry Pool Falls, 200, 204–206, 216
Quebec, xxvii

R
Railroad Notch, 173
Rainbow Creek, 84
Rainbow Falls (AMR), 84–88
Rainbow Falls (Ausable Chasm), xxiv, 26, 86, 147, 154, 156, 158–162
Rainbow Falls (Grasse River), 87
Rainbow Falls (High Falls Gorge), 87, 211
Rainbow Falls (Minnewaska), 87
Rainbow Falls (Oswegatchie River), 87

Rainbow Falls (Watkins Glen), 87
Ralph, Alexander, 185
"random scoots", xxvii
Randorf, Gary, 87, 139, 149, 223
Ranney Bridge, 115
Raquette Falls, 131
Redfield, Mount, 186
Redfield, William C., 186
Reines, Sean, 158
Reiter, Cliff, 12, 96, 133, 179
Resser, Dr. Charles, 157
Resting Rock, 130
Revolutionary War, 166
Rice, Amos, 25
Rice's Falls, 23
Rich, 23
Richman, Jonah, 4
Riverfront Park, 191
"River Rebounds, A", 6
Roadside Falls, 112–114
Roadside Geology of New York, 199, 211
ROANKA Attractions Corp., 211
Roaring Brook, 70
Roaring Brook Falls, xxiv, xxviii, 59, 65, 70–73
Rockachusetts: An Explorer's Guide to Amazing Boulders of Massachusetts, xxxv, 202
Rock & River, 166
rock climbing, 59, 67
Rock Cut Brook, 138, 139
Rock Cut Brook Falls, 141
Rock Garden Falls, 58, 59, 64
Rock of Gibraltar, 2
Rocks and Routes of the North Country, New York, 72, 210
Rocky Falls, 176, 177
Rocky Peak Ridge, 51, 56, 60
Roger, Platt, 35
Rogers Company, J. & J., 134, 180

Roosevelt, Governor Franklin
   D., 225
Rooster Comb Brook, 138, 140
Rooster Comb Mountain,
   110–112, 114–116
Roscoe, John B., 25
Roscoe, Stephen, 25
Roseberry, C. R., 210
Round Mountain, 50, 54, 59, 65,
   77
Round Pond, 11, 16, 52, 56
Route 73 Flume, 17
Rowland, Tim, 119
Rushing Brook Way, 114
Russell Falls, 90, 91

S
Sacandaga River, 44
Sach, Ernest, Dr., 114
Sach's Flume Brook, 114
Saddleback Brook, 136
Saddleback Mountain, 75, 88
"Safety First at Waterfalls", 5
St. Huberts, xxiv, 69, 74, 76
St. Huberts Inn, 75
St. Lawrence River, xxxi, 184
Salmon Falls, xxviii
Sandy, Tropical Storm, 74
Santa's Workshop, 221
Satin, Jordan, 4
Saunders Mountain, 45
Sawteeth Mountain, 88
Sawyer & Mead, 39
Schaefer, Paul, xxviii
Schatz, Scherelene L., 26, 122
Schoharie County, xxv
Schroon River, 38, 41, 42, 44, 48
Schryver, David, 87, 150
Schwartz, Carl, 128
Schwartz Trail, 227–229
Scott Pond, 176
Secluded Cascade, 52, 53
Seneca Ray Stoddard:
   Transforming the Adirondack
   Wilderness in Texts and Image, 159

Sentinel Range, 167
Seventh Annual Report of the
   Progress of the Topographic Survey of
   the Adirondack Region of New York, 95
Shadow Rock Pond, 208
Shanty Brook, 76, 88
Shanty Brook Falls, 88
Sharp Bridge, 44
Shawangunks, xxv, 87
Sherburne, Nathaniel, 125
Sherman Iron Ore Operation, 29
Shoebox Falls, 11–13, 16, 21
Shurtleft, Roswell, xxxi, 76
Silver Cascade, 26, 27
Silver Cascade Brook, 26, 27
Singing Waters, 26
Single X Cave, xxv
Skagrterack Mountain, 163
Slide Brook, 24, 146
Slide Mountain Brook, 132, 133
Slide Mountain Brook Falls,
   132, 133
sluiceway, 24
Smathers, Edward M., 87, 150,
   230
Smith, Clyde, xxviii, 86, 87,
   182–184
Smith, H. P., 39
Snow Mountain, 75, 110, 112,
   114–116
Snow Mountain Brook, 114
"Sound of Falling Water, A",
   xxviii
South Dix, 2
South Fork of the Boquet, Falls
   on the, 30, 31
South Meadow, 173
South Meadow Brook, 175, 199
Southside Trail, 128, 137–146
Spirit of the Adirondacks, 207
Split Falls, 96, 97
Split Rock Falls, 3–8, 123
Split Rock Inn, 6
Split Rock Mountain, 5
"sports", xxxi

Spotted Mountain, 44
Sprakers, 63
Spread Eagle Mountain, 105
Sprucemill Brook, 148
Squires, Dennis, 122, 128
Stag Brook, 216–20
Stag Brook Falls, 216–220
Staircase Falls, 14
Stanley Falls, 161
Stansfield, Dean S., 192
Starmer, Tim, 87
Stechschulte, Ben, 11, 150
Steele, Jonathan, 25
stereoviews, xxxii
Stevens, Hubert, 225
Stillman, William, 152
Stoddard, Seneca Ray, xxxi, 5, 6, 26, 72, 85, 90, 117, 159
Stoneleigh B & B, 27
Street Mountain, 176, 191
Styles Brook, 148
Styles Brook Falls, 148, 149
Styles Brook Falls Bridge, 148
Suicide Leap, 3
Sunderland Cave, xxxv
Surprise Falls, 96
Swedberg, James, 179
Sweeting, Bobbie, 230
"Switzerland of America", xxiv, 104

T

Tahawus, 179
Tefft, Tim, 170
Tenderfoot Falls, 142
Tenderfoot Pool Falls, 142
Tenth Mountain Division of Adirondack Soldiers, 204
Thruway, NYS, 64
Tierney, Marjorie Ann, 111
T-Lake Falls, 70
Town Ridge Loop Trail Falls, 119, 120
Townsend, Solomon, 152
Trails End Inn, xxiv, 111

Trails of the High Peaks Region, 21
Trap Dike, 185–187
Trap Dike Falls, 185–187
Troop, George, 35
Troy, 35
Truesdale, Hardie, 72, 211
Twin Bridges, 25
Twin Falls, 56–58
Twin Pond, 11, 12, 16, 54
Twin Pond Cascade, 54–56
Twin Pond Outlet Stream, 16
Two Adirondack Hamlets in History: Keene and Keene Valley, 75, 77, 170
Two Guides, 119

U

Underwood, 7, 50
Underwood Falls, 7, 8
Uneven Ground, 185
Union Falls Pond, 226
Upper Boquet Falls, 13, 14, 16, 21
Upper Gill Brook Falls, 99–101
Upper Johns Brook Flume & Falls, 131, 132
Up the Lake Road, 83, 88, 94
US, 23
US Gorge, 23, 24
US Mountain, 23

V

Van Diver, Bradford B., 72, 125, 199, 210, 211
Van Hoevenberg, Henry, 171, 172, 175, 176, 180, 182
Vermont, 4
Vermont Waterfalls: A Guide, 24
"Violence in the Valley", 74
Virginia Creek, 40
Viscome, Laura, 4

W

Wadhams, 28, 31

Wadhams Falls, 28–32
Wadhams Free Library, 30
Wadhams, General Luman, 30, 31
Wadhams Hydro Electric Station, 29
Wadhams Mills, 31
Wagon Wheel Falls, 39
Wagon Wheel Landing, 40
Wallace, Edwin. R., 117, 180
Wallace Falls, 180
Wall of Jericho, 209
Walton Bridge, 125
Wanika Falls, xxviii, 191, 193–198
Warden's Camp, 88
Warner, Charles Dudley, 101, 105, 117
War of 1812, 2
Washbond Flume Brook, 114
Washbond, Henry, 114
Washbond's Flume, 114
Waterfalls of New York State, 87, 150
Waterfalls of the Adirondacks and Catskills, 8, 17, 31, 83, 149, 207, 212
Waterfall Weekend, xxiv, 31, 105
waterpower, xxxii
Watkins Glen, 87
Watson, Winslow C., 76, 84, 159, 162, 168, 200
Weber, Sandra, 180, 228
Wedge Brook, 80
Wedge Brook Falls, 80–82, 90
Wells, 44
Wells Dam, 74
Wells, Sylvanus, 76, 125
West Branch Nature Trail, 216
West Inlet, 39, 40
West Inlet Falls, 39
West Mill Brook, 41, 43
Weston, Warren F., 169
Westport, 29
West River Trail, 78–88

Whiteface Access Road Bridge, 214
Whiteface Brook, 200, 202, 203
Whiteface Brook Falls, 200–204
Whiteface Landing, 200, 202, 203
Whiteface Mountain, 200, 202, 203, 207, 216, 221, 225, 226, 228
Whiteface Mountain Ski Center, 213–215, 220, 224
Whiteface Veteran's Memorial Highway, 225
Whiteface Veteran's Memorial Highway Cascade, 225, 226
White, William Chapman., 226
Williams, C. N., 25
Williams, Colonel Edward F., 23
Willsboro, 2, 35
Willsboro Falls, 34–36
Willsboro Pulp Mill, 35
Wilmington, 200, 211, 221
Wilmington and the American Mind, xxxi
Wilmington and the Whiteface Region: Images of America, 211, 222
Wilmington Flume, 221
Wilmington Historical Society, 211, 222
Wilmington Notch, 199, 200, 209
Wilmington Notch Campground, 212, 215, 216
Wilmington Notch Falls, 200, 212, 213, 215, 216
Winkler, John, 20, 43, 101, 102
Winter Camp, 134
Winter Olympics, 214
Withersbee, 29
Witness the Forever Wild: A Guide to Favorite Hikes Around the Adirondack High Peaks, 12, 96, 133, 179
Wolf Jaw Brook, 138, 139, 144
Wolf Jaw Brook Falls, 144–146

Wolfjaw, Lower, 75, 79, 80, 144
Wolfjaw, Upper, 75, 82
Wood, Charles M., 25
"Workhorses of the Industrial Revolution", xxxii
World War II, 210
Wrap-around Falls, 218
Wright Peak, 178, 180
Wuerthner, George, 125, 167, 180
Wyant, A. H., xxxi, 76
Wyant, Alexander, 144
Wyckoff, Jarome, 87, 185

Y

Yard Mountain, 132, 173
York State Traditions, 157
"Yosemite in Miniature", 157
"Yosemite of the East", 104

Z

Zander Scott Trail, 58, 60, 64